WILLIAM SMITH

THE SCIENCE OF NOTHINGNESS

How Empty Space Became the Most Powerful Force in the Universe

First edition

This book was professionally typeset on Reedsy.
Find out more at reedsy.com

Contents

Prologue 1
Introduction 3
1 The Lie We All Believe 6
2 When Mathematics Discovered Nothing 11
3 The Experiment That Changed Everything 16
4 What Vacuum Really Means 21
5 The Uncertainty That Creates Reality 26
6 Particles From Absolutely Nowhere 31
7 The Force of Emptiness 36
8 The Energy You Cannot Remove 41
9 Why Reality Doesn't Collapse 46
10 The Biggest Discovery Since Gravity 52
11 Dark Energy—The Weight of the Void 58
12 How Nothing Pushes 64
13 Before the Beginning 69
14 The Quantum Origin 74
15 The Vacuum That Could End Us 79
16 The Hardest Question 83
17 Is Nothing More Natural Than Something? 89
18 Why Our Brains Fail at Nothing 94
19 What Nothingness Teaches Us About Everything 99
20 EPILOGUE 104

Prologue

THE ROOM BETWEEN EVERYTHING

The Experiment

Put this book down for a moment. Look around you. Whether you are sitting in a quiet library, a crowded coffee shop, or your own living room, I want you to perform a mental act of destruction. I want you to empty the room.

Start with the obvious things. Mentally remove the furniture. Banish the chairs, the tables, the lamps, and the rugs. Send the other people away. Remove the walls, the floor, and the ceiling, so that you are floating in an open, undefined expanse. Now, go further. Siphon out the air. Remove every molecule of nitrogen, oxygen, and carbon dioxide until there is no wind, no pressure, and no sound.

Now, turn out the lights. Banish every photon from the sun or the streetlamps. Shield this space from the cosmic rays raining down from the galaxy and the radio waves passing through from local cell towers. Cool the temperature down. Slower. Colder. Drop it to absolute zero, so that even the thermal jitter of heat ceases to exist.

Finally, sweep away the dust. Find the stray hydrogen atoms floating in the interstellar void and kick them out. Ensure there is absolutely no matter, no radiation, no heat, and no light.

What is left?

The Active Silence

Common sense tells you that you are now looking at "nothing." You are looking at a void—a passive, empty stage where reality used to be. It is dark. It is silent. It is dead.

Common sense is wrong.

If you could look at this "empty" space with the eyes of a physicist, you would not see a blank canvas. You would see a catastrophe of energy. In the silence, you would hear the roar of quantum fields vibrating with the frantic energy of creation. You would see "virtual particles" ghostly pairs of matter and antimatter popping into existence and annihilating each other in trillionths of a nanosecond, a seething foam of reality that never sits still.

You would feel weight. This empty space is not weightless; it carries a heavy, invisible energy that pushes the universe apart, a force so powerful it is currently tearing galaxies away from one another. And if you looked close enough down to the smallest possible scale you might see the geometry of space itself twisting, knotting, and foaming like boiling water.

This "nothing" is not the absence of the story. It is the main character. It is the source of the Big Bang, the glue that holds atoms together, and the eventual destroyer of the cosmos. We have spent thousands of years obsessed with the "stuff" of the universe—the stars, the planets, the people. But the stuff is just the garnish. The universe is 99.9% void, and it is time we met the void face to face.

Introduction

THE INVITATION

What You Are Holding in Your Hands

This book is an exploration of the most difficult concept in human history. It is a biography of a ghost. "Nothingness" is an idea that has driven mathematicians mad, terrified religious authorities, and ultimately forced physicists to rewrite the laws of nature. It is a concept that feels intuitive we all know what it means for a box to be empty but as soon as we poke it, it falls apart.

I wrote this book because I became obsessed with a single, nagging paradox: **How can "nothing" be the most powerful force in the universe?**

We are taught from a young age that the world consists of things. We learn to count apples, block with toy bricks, and avoid bumping into walls. Our brains are hardwired to process objects. But the deeper we delve into the nature of reality, the more we realize that "objects" are an illusion. The solid world is a temporary fluctuation in a vast, ocean-like vacuum. To understand the universe, we must stop looking at the waves and start looking at the water.

How to Read This Book

This is a science book, but it is also a story. It is the story of how humanity discovered that the empty space between the stars is alive.

- **Part I: The Illusion of Emptiness** takes you through the history of the void, from the Greek philosophers who feared it to the 19th-century scientists who tried to fill it with "aether," ending with the shocking revelation that atoms are mostly empty space.
- **Part II: The Quantum Void** dives into the microscopic world, where we discover that "zero" is an impossible number and that the vacuum is teeming with temporary, ghostly particles.
- **Part III: Nothing on a Cosmic Scale** zooms out to the edge of the universe, where we confront Dark Energy—the mysterious force of nothingness that makes up 68% of reality and dictates our cosmic destiny.
- **Part IV: When Physics Meets Meaning** tackles the philosophical vertigo that comes with these discoveries. If the universe emerged from nothing, what does that mean for existence? Why is there something rather than nothing?

A Warning

You do not need a PhD in physics to read this book. I have avoided dense equations in favor of metaphors, diagrams, and narratives. However, you do need an open mind. The concepts we will discuss are deeply counter-intuitive. They violate the common sense you use to navigate your daily life.

You will learn that empty space has weight. You will learn that energy can be borrowed from nowhere. You will learn that the universe might be a bubble that could pop at any moment.

If you feel confused at times, that is a good sign. As the physicist Niels Bohr once said, "Anyone who is not shocked by quantum theory has not understood it." The feeling of confusion is simply the sensation of your brain trying to resize itself to fit a reality larger than it was evolved to comprehend.

So, take a deep breath. Let go of your certainty. Let's step into the void together.

One

The Lie We All Believe

The Space Between the Stars

Sometime in your childhood, likely on a clear night away from the city lights, you looked up and asked a question that has plagued humanity for thousands of years. You pointed to the dark gaps between the bright pinpricks of stars and asked, "What is that?" The answer you received was almost certainly immediate and confident: "That is space." If you pressed further and asked what space actually *is*, the answer would have been even simpler: "It's nothing."

We grow up accepting this explanation as an absolute truth. We view the universe as a vast, dark stage where the drama of existence plays out, but we assume the stage itself is merely a backdrop passive, empty, and devoid of properties. We intuitively divide reality into two distinct categories: "stuff" and "emptiness." There are planets, stars, trees, and people, and then there is the empty container that holds them all. This distinction feels natural because our brains evolved to survive on the African savanna, where identifying a lion (something) hidden in the grass (something else) was a matter of life and death, while contemplating the nature of the empty air between them was biologically useless.

However, this intuition is a lie. It is perhaps the single greatest misconception we hold about the nature of reality. To say that the space between stars or even the space between your eyes and this page is "nothing" is to ignore one of the most vibrant, heavy, and energetic entities in the universe..But before we can understand what this "nothing" actually is, we must first dismantle our deep-seated belief that emptiness is the default state of the universe. We have to look at the solid world around us and realize that it is a ghost.

The Horror of the Vacuum

Our aversion to the concept of true nothingness is not just a modern habit; it is a deep philosophical anxiety that stretches back to antiquity. The ancient Greeks were obsessed with the nature of the void, but they largely refused to believe it could exist. Aristotle, whose ideas dominated Western thought for nearly two millennia, argued vehemently against the existence of a true vacuum. He proposed the concept of *horror vacui,* or "nature abhors a vacuum." To Aristotle, space was a continuum of matter; if you moved a rock, air immediately rushed in to fill the gap. The idea of a spot in the universe containing absolutely nothing was considered logically impossible and physically repugnant.

This fear of the void stalled scientific progress for centuries. It wasn't just a physical claim; it was a theological one. If God is everywhere, how could there be a place where nothing exists? It implied a limit to divine power. Consequently, early scientists invented invisible fluids to fill the gaps in their theories. When they couldn't explain how light traveled from the sun to the earth, they invented the "coniferous ether," a subtle substance that supposedly permeated the entire universe, simply because they could not accept that light could travel through nothing. We have always tried to stuff the void with imaginary matter because the alternative that the universe is mostly empty was too terrifying to confront.

The Cathedral of Emptiness

The illusion of solidity began to crack in the early 20th century, not by looking at the stars, but by looking inward at the fundamental building blocks of matter. For a long time, we believed atoms were solid little billiard balls indivisible chunks of matter packed together to create the world. But when physicist Ernest Rutherford began firing alpha particles at thin sheets of gold foil in 1911, he discovered something shocking. Most of the particles passed straight through the gold as if it weren't there. Only a tiny fraction bounced back.

Rutherford had discovered the atomic nucleus, but in doing so, he inadvertently revealed the ghostly nature of our reality. The picture of the atom that emerged from his work, and later refined by Niels Bohr and modern quantum mechanics, is disturbing. An atom is not a solid ball; it is a tiny, dense nucleus surrounded by a vast cloud of probability where electrons might be found.

To visualize the scale of this emptiness, we need to expand our perspective. Imagine an atom expanded to the size of a professional football stadium. In this analogy, the nucleus where 99.9% of the atom's mass resides would be a marble sitting on the 50-yard line. The electrons? They would be tiny gnats buzzing in the highest seats of the nosebleed section. And what is between the marble and the gnats? What fills the rest of the stadium?

Absolutely nothing.

This means that everything you regard as solid is a deception. The chair you are sitting on, the floor beneath your feet, and the hand holding this book are 99.9999999% empty space. If you were to remove all the empty space from the atoms that make up the entire human race—all eight billion of us—and compress us down to just our physical matter, we would all fit inside the volume of a single sugar cube. That sugar cube would weigh incredibly heavy, roughly as much as a mountain, but it would reveal the stark truth: we are mostly scattered dust suspended in a void.

The Illusion of Touch

If we are mostly empty space, why don't we fall through the floor? Why does a rock feel hard? When you clap your hands, why do they stop with a sharp *smack* rather than passing through one another like ghosts?

The answer lies in the forces that govern the void, not in the matter itself. You have never actually touched anything in your entire life. When you press your finger against a table, the atoms of your finger never come into physical contact with the atoms of the table. As your finger approaches the surface, the electromagnetic fields generated by the electrons in your finger repel the electromagnetic fields of the electrons in the table. The sensation of "solidity" or "pressure" is actually your brain interpreting the resistance of these magnetic fields pushing against each other.

You are hovering, perpetually, on a cushion of electromagnetic force, separated from the rest of the world by an atomic abyss. The solidity of the world is a macroscopic illusion created by the interplay of forces across empty space. We are not solid objects moving through a void; we are mostly void moving through void, held together by invisible rules.

Facing the Void

This revelation is the starting point of our journey. We must abandon the Aristotelian comfort that the world is full and solid. We have to accept that "matter" is a rare impurity in a universe of emptiness. But here is the twist that will drive the rest of this book: just because that space is empty of *matter* does not mean it is empty of *reality*.

As we stripped away the air, then the light, and finally the atoms, we seemed to be left with nothing. But as we will discover in the coming chapters, this "nothing" is deceptively alive. The void that makes up 99.9% of your body and the vast gaps between the galaxies is not a passive stage. It is a frothing, boiling, expanding ocean of energy that dictates the fate of the universe. We used to think that matter was the main character of the cosmos and space was just the stage. We were wrong. The matter is just the props; the empty space is where the real action is.

To understand this, however, we must first understand how humanity learned to talk about nothing. Before we could discover the physics of the vacuum, we had to invent the mathematics of zero. And as we will see, the struggle to accept zero was just as difficult, and dangerous, as the struggle to accept the physical void.

Two

When Mathematics Discovered Nothing

The Number That Wasn't There

It is difficult for the modern mind, which is steeped in a digital world built entirely on a binary code of ones and zeros, to comprehend that for the vast majority of human history, the concept of zero did not exist. Civilization managed to rise, build empires, construct the Pyramids of Giza, and map the movement of the stars without ever writing down a symbol for "nothing." To the ancient world, numbers were tools for counting tangible realities cattle, soldiers, bushels of wheat, or coins in a treasury and it made absolutely no sense to assign a symbol to a lack of those things. If a farmer had no sheep, he did not need a number to describe his situation; he simply had no sheep. The absence was self-evident and required no mathematical notation.

This absence of zero was not merely an oversight; it was a psychological block rooted in the way early humans perceived the world. The Greeks, despite their profound contributions to geometry and logic, had a deep philosophical aversion to the concept of the void. Pythagoras and his followers believed that the universe was ruled by harmony and ratios, and "nothing" was the antithesis of existence a chaotic abyss that had no place in the perfect order of numbers. To them, numbers were geometric shapes: one was a point, two was a line, three was a triangle, and four was a pyramid. You could not draw a shape for nothing. Consequently, the greatest mathematicians of antiquity performed complex calculations using clunky, alphanumeric systems like Roman numerals, where performing simple multiplication or division was a Herculean mental task. Try, for a moment, to divide MCCXLVIII by XII without converting them to modern numbers, and you will understand the stranglehold that the lack of zero placed on human progress.

The Gift from the East

The revolution that would eventually unlock the secrets of the universe did not begin in Greece or Rome, but in India, around the 5th century AD. While the West was terrified of the void, Eastern philosophy embraced it. In the spiritual traditions of Buddhism and Hinduism, the concept of emptiness *Sunya* or *Shunyata* was not a frightening abyss but a state of potential and liberation, the ultimate goal of spiritual practice. It was within this cultural framework that the mathematician Brahmagupta finally gave nothingness a name and a shape. He defined zero not merely as a placeholder a way to distinguish 10 from 1 but as a number in its own right, with its own specific rules and properties.

Brahmagupta wrote the first rules of arithmetic for zero, stating that if you subtract a number from itself, you get *Sunya*. This seems trivial to us today, but at the time, it was a radical abstraction that severed the link between mathematics and physical objects. He went further, establishing that zero interacts with other numbers: adding zero changes nothing, multiplying by zero collapses everything into nothing. This was the moment humanity invented a symbolic language for the void. The Indians represented this concept with a small dot, which eventually evolved into the hollow circle we use today—a literal hole on the page representing the hole in reality. This new number system, which included zero, eventually traveled to the Islamic world, where the great Persian mathematician AL-Khwarizmi synthesized Indian arithmetic into what we now call algebra. The word "algorithm" comes from his name, and the word "zero" is derived from the Arabic *sifr*, meaning "empty."

The Devil's Doorway

When these "Arabic numerals" finally reached Europe in the Middle Ages, they were not welcomed as a scientific breakthrough; they were greeted with suspicion, fear, and hostility. The Catholic Church, which held dominion over European intellectual life, found the concept of zero to be theologically perilous. In their worldview, God was infinite, the "Alpha and Omega," the creator of all things. To assert the existence of "nothing" was to imply a place where God was not, which was tantamount to atheism or the presence of the Devil. The void was the realm of Satan, and zero was his number.

Furthermore, the new number system was viewed as a tool of deception. In the old Roman system, it was very difficult to forge numbers; a "V" (5) could not easily be turned into an "X" (10). But with the Arabic system, a merchant could easily add a zero to the end of a ledger entry, turning 10 coins into 100 with the stroke of a pen. Because of this potential for fraud and its foreign, "heathen" origins, the city of Florence banned the use of Arabic numerals in 1299. The word "cipher," which originally meant zero, came to mean a secret code, because those who wished to use this superior mathematical system had to do so in secret, hiding their calculations from the authorities. It took centuries for the utility of zero to outweigh the dogma of the Church, a transition that only truly solidified with the rise of banking and international trade, where the efficiency of zero was worth more than philosophical comfort.

Dividing by Nothing

Once zero was accepted, it immediately began to cause problems, the most significant of which was the catastrophe of division. In mathematics, division is essentially the question: "How many times does this number fit into that number?" If you divide 10 by 2, you are asking how many groups of 2 fit into 10. The answer is 5. But if you try to divide 10 by zero, you are asking, "How many times does nothing fit into 10?" You can add nothing to itself a billion times, or a trillion times, and you will never reach 10. The answer is not a number; it is an explosion of logic.

As the denominator in a fraction gets smaller and smaller, approaching zero, the result gets larger and larger. If you divide 1 by 0.1, you get 10. Divide 1 by 0.000001, and you get a million. As you slide closer to the edge of the abyss, the value shoots upward toward infinity. In this way, zero and infinity are revealed to be twin reflections of one another two sides of the same coin. Zero is the infinitely small, and its reciprocal is the infinitely large. This relationship is not just a mathematical curiosity; it is the reason why the laws of physics break down at the center of a black hole. When we try to calculate gravity at a single point (zero volume), our equations result in "singularities" mathematical errors where values become infinite and reality ceases to make sense. Zero is the monster that lives in the heart of our equations, threatening to destroy them if we are not careful.

The Calculus of the Void

The true power of zero was finally harnessed in the 17th century by Isaac Newton and Gottfried Wilhelm Leibniz, who used it to invent calculus. They realized that while you cannot divide by zero, you can get *infinitely close* to it. They developed a method of dealing with "infinitesimals" quantities that are astonishingly small, hovering just above the void of zero without quite falling in. By dancing on the edge of nothingness, they were able to calculate the slope of a curve at a single point and the motion of planets in orbit.

Calculus is essentially the science of slicing reality into pieces that are almost zero in size, and then adding them all back up. Without this ability to manipulate the concept of "almost nothing," we would have no modern physics, no engineering, no space travel, and no understanding of electromagnetic fields. We had to master the concept of the mathematical void before we could hope to understand the physical one.

However, a dangerous paradox remained. In mathematics, zero is a pristine, perfect concept it represents absolute nullity, a complete absence of value. $1 - 1 = 0$ is a clean, absolute truth. But as we began to apply this mathematics to the physical universe, we encountered a jarring problem: nature does not seem to allow "true" zeros. In the abstract world of the mind, you can have nothing. In the physical world of quantum mechanics, as we are about to see, "nothing" is a forbidden state. We built a beautiful mathematical language based on the existence of zero, only to discover that the universe itself might reject the very concept we worked so hard to invent.

Three

The Experiment That Changed Everything

The Ocean of Light

By the late 19th century, the edifice of physics seemed nearly complete. Isaac Newton's laws of motion ruled the mechanical world, and James Clerk Maxwell's elegant equations governed electricity and magnetism. Scientists felt a confident swagger, believing they were just a few decimal points away from understanding the entire universe. But there was one nagging loose end, a subtle contradiction that centered on the nature of light itself.

For decades, a fierce debate had raged: was light a particle (a stream of tiny bullets) or a wave (a ripple of energy)? By the 1800s, the wave theory had won decisively. Experiments showed that light could bend around corners and interfere with itself, behaviors that were undeniably wave-like. But this victory created a massive logical hole. If light was a wave, what was it waving *through?*

Sound waves need air; ocean waves need water; seismic waves need the earth.

You cannot have a wave without a medium a physical substance for the energy to travel through. If you ring a bell in a vacuum jar where all the air has been removed, the sound vanishes. Yet, sunlight travels through 93 million miles of empty space to reach Earth. The stars send their light across billions of light-years of void. If space were truly empty, how could the waves of light possibly traverse it? It was as absurd as watching ocean waves crash onto a beach where there was no ocean.

The Ghostly Wind

To solve this problem, 19th-century scientists posited the existence of the "luminiferous aether" (or ether). This was not the chemical ether used to knock out patients in surgery; it was a hypothetical, invisible, weightless substance that supposedly filled every cubic inch of the universe. It was the medium of light. The aether had to be everywhere permeating the vacuum of space, the atmosphere of Earth, and even the spaces between atoms in your body. It had to be incredibly rigid to allow light to travel so fast (just as sound travels faster in steel than in air), yet it had to be completely friction less so that planets could orbit through it without slowing down.

It was a magical, paradoxical substance, but physicists were convinced it had to exist. If there was an aether, then the Earth must be moving through it as it orbits the sun. Imagine running through still air; you feel a wind hitting your face. Similarly, as the Earth plows through the stationary aether at 67,000 miles per hour, we should experience an "aether wind." Detecting this wind became the holy grail of physics. If we could measure it, we would prove the existence of the background fabric of the universe.

The Most Famous Failure in History

Enter Albert Michelson and Edward Morley. In 1887, at what is now Case Western Reserve University in Ohio, these two scientists set out to measure the aether wind. Michelson, a perfectionist obsessed with precision, designed an instrument of breathtaking sensitivity called the interferometer.

The logic of the experiment was elegant and undeniable. Imagine two swimmers in a river. One swims across the river and back; the other swims upstream and then back downstream. Even if they swim the same distance at the same speed, the math shows that the swimmer fighting the current (upstream) and then riding it (downstream) will always take longer than the one cutting across the current. The current slows you down more than it helps you speed up.

Michelson and Morley built a device to race beams of light instead of swimmers. They split a beam of light into two paths at right angles to each other. One beam was sent in the direction of the Earth's motion (into the aether wind), and the other was sent perpendicular to it (across the wind). Mirrors reflected the beams back to a central point where they recombined.

If the aether existed, the beam fighting the "wind" should have been slightly delayed compared to the other. When the two beams recombined, their light waves would be out of sync, creating a characteristic pattern of interference fringes—bands of light and dark. Michelson floated his massive device on a pool of mercury to eliminate vibrations from passing street traffic and rotated it slowly, expecting to see the interference pattern shift as the arms aligned with the aether wind.

The Sound of Silence

They saw... nothing.

The interference pattern did not shift. The light beams arrived at exactly the same time, regardless of which way the Earth was moving. It was as if the Earth was standing perfectly still in the center of the universe which they knew was impossible or as if the aether wind simply didn't exist.

They were baffled. They assumed they had made a mistake. They checked the mirrors, the angles, the light source. They repeated the experiment at different times of day and different seasons of the year, thinking perhaps the Earth's position in orbit mattered. Every single time, the result was the same: null. Zero.

This "failed" experiment sent shock waves through the scientific community. It was a catastrophe. The most trusted theory of the 19th century had just collided with reality, and reality had refused to budge. Scientists twisted themselves into knots trying to save the aether. Some suggested the Earth "dragged" the aether along with it like a sticky atmosphere. Others, like Henrik Lorentz, proposed that objects physically shrank in the direction of motion just enough to hide the effect. But these were desperate patches on a sinking ship.

Einstein's Revolution

It took a patent clerk in Bern, Switzerland, to look at the wreckage of the aether theory and see the truth. In 1905, Albert Einstein published his theory of Special Relativity, and he did something radical: he simply deleted the aether.

Einstein asked a dangerous question: "What if light *doesn't* need a medium?"

He proposed that light is a self-propagating wave a bundle of electric and magnetic fields that sustain each other, requiring no physical substance to wave through. He declared that the speed of light is constant for all observers, regardless of their motion. In doing so, he removed the need for the background stage.

This was a profound conceptual earthquake. Before Einstein, "nothing" (space) was thought to be "something" (aether). Einstein stripped the universe of this invisible fluid and returned us to the true void. Space, he argued, is not a substance; it is a geometry. It is a relational framework where events happen, but it is not made of "stuff" in the way water or air is.

The Vacuum Returns

The Michelson-Morley experiment is often called the most successful failed experiment in history because, by failing to find what they were looking for, they found something far more important. They found the modern vacuum.

However, Einstein's victory created a new, deeper mystery. If space isn't filled with aether, then what *is* it? Is it truly just an abstract coordinate system, a mathematical grid with no physical reality? For a few decades, it seemed so. Physics became comfortable with the idea of a truly empty void where geometry ruled.

But this comfort was temporary. While Einstein was cleaning up the universe on the largest scales, a new breed of physicists was beginning to look at the smallest scales—the quantum realm. And they were about to discover that while the ether was dead, the vacuum was far from empty. Einstein had emptied the room, but quantum mechanics was about to fill it back up with ghosts that were far stranger than any 19th-century invisible wind. The void was about to wake up.

Four

What Vacuum Really Means

The Linguistic Trap

In our daily lives, the word "vacuum" is a mundane utility. It brings to mind household appliances, sealed coffee bricks, or perhaps the light bulb filaments of the 20th century. To a layperson, a vacuum is simply a volume of space where the air has been sucked out. It is defined negatively: a vacuum is a place where something is missing. If you drink a milkshake through a straw, you are using a partial vacuum to pull the liquid up. If you pump the air out of a bell jar, the sound of a ringing alarm clock inside will fade to silence. In this classical view, a vacuum is a passive, quiet state a "lesser" version of reality achieved by subtraction.

However, to a physicist, the word "vacuum" carries a much heavier, more specific, and profoundly stranger meaning. In the lexicon of modern science, a vacuum is not simply "emptiness." It is a physical state defined by a specific energy configuration. The gap between the dictionary definition of vacuum ("a space entirely devoid of matter") and the physical definition ("the lowest possible energy state of a system") is the chasm where our common sense falls to its death. We must stop thinking of the vacuum as a "hole" in the universe and start thinking of it as the "ground floor" of the universe. It is the foundation upon which the house of reality is built, and as any architect knows, the foundation is often the most critical part of the structure.

The Hierarchy of Emptiness

To understand what a true vacuum is, we must perform a mental experiment of extreme subtraction. Imagine we have a box made of infinitely strong steel, one cubic meter in size, sitting on a table in a laboratory. Our goal is to create a perfect void inside this box. We want to remove absolutely everything until nothing remains.

Level 1: The Mechanical Vacuum

First, we attach a powerful industrial pump to the box and suck out the air. We remove the nitrogen, the oxygen, the argon, and the carbon dioxide. This is what we call a "rough vacuum." It is empty enough for industrial manufacturing or freeze-drying food, but it is still teeming with matter. Even in a high-quality laboratory vacuum, there are still billions of gas molecules zipping around inside that cubic meter, bouncing off the walls.

Level 2: The Interstellar Vacuum

We upgrade our technology to the most advanced turbo-molecular pumps available, or perhaps we transport our box to the space between the stars. Here, the density of matter drops precipitously. In the vast molecular clouds of the galaxy, there might be only a few thousand atoms per cubic centimeter. In the voids between galaxies, it might be just one atom per cubic meter. But even if we manage to chase down every single stray hydrogen atom and eject it from our box, we are not done.

Level 3: The Thermal Vacuum

Our box is now chemically empty there are no atoms left inside. But it is not truly empty. If the walls of the box are at room temperature, they are radiating heat. Infrared photons particles of light are streaming from the walls, crisscrossing the interior of the box at the speed of light. The box is full of a "photon gas." To get rid of this, we must cool the box. We freeze it down to absolute zero (-273.15°C or 0 Kelvin). As the temperature drops, the radiation from the walls fades. The frantic energy quiets down.

Level 4: The Cosmic Echo

But we still haven't succeeded. The universe itself is permeated by the Cosmic Microwave Background (CMB) the afterglow of the Big Bang. These ancient photons are everywhere, passing through the Earth, your body, and the walls of our steel box (unless we shield it with superconducting lead). To create a true void, we must shield the box from every source of radiation in the universe: radio waves, neutrinos, cosmic rays, and gravity waves.

Let us assume, for the sake of argument, that we can do the impossible. We have a box with no atoms, no heat, no light, no neutrinos, and walls at absolute zero. We have stripped away every distinct particle that physics knows how to name. What is left?

The Indestructible Fabric

According to classical intuition, the box should now be truly empty. It should be a region of pure "nothing." But quantum mechanics tells us that if you look inside this perfectly shielded, frozen, evacuated box, you will not find stillness. You will find a distinct, measurable presence.

The reason is that you cannot remove the *fields*.

In modern physics, the fundamental ingredients of the universe are not particles; they are fields. A field is a fluid-like substance that permeates all of space and time. There is an electromagnetic field (which gives us light and magnetism), an electron field (which gives us electricity and matter), a Higgs field (which gives mass), and several others. These fields are like the water in an ocean; they are everywhere, even if there are no waves.

A "particle" like an electron is simply a ripple or a knot in the electron field, just as a wave is a disturbance in the ocean. When we sucked the atoms out of our box, all we did was calm the waves. We removed the excitation. But the ocean itself the fields remained. You cannot scoop the fields out of the box because the fields *are* the space inside the box. To remove the fields would be to remove space itself, collapsing the box into non-existence.

The Ground State

This leads us to the modern definition of the vacuum: **The vacuum is the state where all fields are as quiet as they can possibly be.**

Note the phrasing: "as quiet as they can possibly be," not "silent." This distinction is crucial. In classical physics, a pendulum that stops swinging hangs perfectly straight and still. It has zero energy. But in quantum mechanics, as we will explore in the next section, absolute stillness is forbidden. A quantum field can never have zero energy. It must always be tittering, vibrating, and fluctuating, even in its lowest possible state.

This "hitter" is inherent to reality. It means that even in our hypothetical, perfect box, the electromagnetic field is still humming with a faint, irreducible energy. This energy is not coming from the outside; it is intrinsic to the vacuum itself. We call this the **Zero-Point Energy**, a term that sounds like science fiction but is a measured, verified fact of nature.

The Basement is Flooded

The implications of this redefinition are staggering. It means that "empty space" is a material with physical properties. It has a structure. It has density. It can be stretched, compressed, and bent. When we look out at the dark night sky, we are not looking at a void; we are looking at a "plenum" a space that is fully occupied by overlapping quantum fields in their ground state.

Imagine walking into a room that appears empty. There is no furniture, no people, no sound. But then you look down and realize the floor is covered in water up to your ankles. You can remove the furniture, but you cannot remove the water because it is part of the room's architecture. That is the quantum vacuum. We live our lives wading through this invisible ocean, oblivious to it because it is uniform and omnipresent. We only notice the fields when they wave (creating light) or splash (creating particles), but the water is there, always.

This realization sets the stage for the greatest drama in physics. If the vacuum is not nothing if it is a seething, vibrating ocean of energy then it can *do* things. It can push plates together. It can glow. It can spawn particles out of thin air. And, as we will see later, it might even be able to blow up the entire universe. The vacuum is not the absence of the story; it is the ink in which the story is written.

Five

The Uncertainty That Creates Reality

The Death of the Clockwork Universe

For centuries, the ultimate goal of science was precision. We believed that the universe was a magnificent, intricate clockwork mechanism, wound up at the beginning of time and ticking forward with predictable, unwavering logic. In this classical view, pioneered by Isaac Newton and polished by Pierre-Simon Laplace, the state of the universe at any given moment was the direct, inevitable result of the moment before. If you knew the position and speed of every particle in the cosmos, and the forces acting upon them, you could in theory predict the future with absolute certainty. There was no room for chance; there was only ignorance of the variables. In this deterministic world, "empty space" was simply the gap between the gears, a silent void where nothing happened until a particle moved through it.

But in the 1920s, a young German physicist named Werner Heisenberg dismantled this clockwork dream and replaced it with something far more unsettling. While working on the new mathematics of quantum mechanics on the island of Heligoland, Heisenberg realized that the universe had a fuzzy bottom limit. He discovered that on the subatomic scale, nature refuses to be pinned down. There is a fundamental graininess to reality, a limit to how precise the universe allows itself to be. This was not a limitation of human technology or clumsy measuring instruments; it was a limitation woven into the very fabric of existence. The implications for "nothingness" would be revolutionary: if you cannot define the state of a system with perfect precision, you can never say with 100% certainty that a box is empty.

The Cosmic Speed Limit of Knowledge

Heisenberg formalized this discovery in his famous Uncertainty Principle. Most popular explanations describe this as a "clumsy observer" problem the idea that to measure a particle, you have to hit it with a photon of light, which knocks it off course, thus changing what you were trying to measure. While true, this explanation is incomplete and slightly misleading. The Uncertainty Principle is much deeper: it states that a particle *does not possess* a precise position and a precise momentum simultaneously. It is not that we don't know where it is; it is that "where it is" is not a single point, but a smear of probability.

The more precisely you try to define one property, the more the universe blurs the other. It is a cosmic seesaw. If you clamp down on a particle to know exactly where it is (zero uncertainty in position), its momentum becomes infinitely uncertain. This means the particle could be moving at any speed, in any direction, effectively exploding out of your grasp. This fuzziness means that the smooth, continuous trajectories of classical physics do not exist at the microscopic level. Particles do not glide like planets; they jump, jitter, and blur. And if particles—the building blocks of matter are subject to this jitter, then the space they occupy must also be subject to it. The "stage" of the universe is not a smooth floor; it is a vibrating trapdoor.

The Energy Heist

The most profound version of Heisenberg's principle for our story, however, is not about position and speed. It is about energy and time. The equation is written as Delta E Delta $t \geq \hbar/2$. In plain English, this equation says that there is a trade-off between how well you know the energy of a system (Delta E) and how long you observe it (Delta t).

If you observe a system for a very long time (large Delta t), you can know its energy very precisely. But if you look at a patch of space for only a tiny fraction of a second (tiny Delta t), the uncertainty in the energy (Delta E) becomes massive.

This mathematical loophole allows the universe to commit a crime: **conservation of energy can be violated, as long as it is done quickly.**

Think of it like a bank loan. In the classical world, you cannot spend money you do not have. Energy is conserved; you cannot create it from nothing. But the quantum bank of the universe has a special rule: you *can* borrow energy from absolutely nowhere, creating it out of thin air, provided you pay it back so fast that no one notices. The more energy you borrow, the faster you have to repay it. You can borrow a tiny amount of energy for a microsecond, or a massive amount of energy for a billionth of a trillionth of a second.

The Seething Vacuum

This "borrowing" mechanism destroys the concept of a static vacuum. Let us return to our empty box from Chapter 4. We removed all the atoms and cooled it to absolute zero. We think the energy inside is zero. But Heisenberg says, "Wait. If you say the energy is *exactly* zero, you are claiming infinite precision. That is forbidden."

Because the duration of time flows on, the uncertainty relation demands that the energy of the vacuum must fluctuate. At any given nanosecond, in any cubic millimeter of empty space, energy is being borrowed from the bank of nothingness. This energy doesn't just sit there; it manifests as matter. Pairs of particles matter and antimatter spontaneously pop into existence, borrowing the energy of their mass from the vacuum. They exist for a fleeting instant, perhaps 10^{-21} seconds, and then they crash back together, annihilating each other and returning their energy to the bank.

We call these **virtual particles**. They are the ghosts of the quantum world. They are not "real" in the sense that they can fly across the universe and hit a detector, but they are not "imaginary" either. They are temporary fluctuations of reality. This means that if you could zoom in on a patch of "empty" space with a microscope powerful enough, you would not see a dark, placid void. You would see a violent, chaotic storm. Particles are popping in and out of existence like the static on a detuned television channel. The vacuum is seething. It is a roiling boil of creation and destruction that never ceases, occurring everywhere in the universe simultaneously, even inside your own body.

The Quantum Foam

In the 1950s, the physicist John Wheeler coined a beautiful term for this phenomenon: **Quantum Foam**. He asked us to imagine flying high above the ocean in an airplane. From 30,000 feet, the ocean looks like a smooth, blue sheet. It appears flat, static, and empty of features. This is how we perceive space-time in our everyday lives smooth and continuous.

But as you descend in a parachute, the view changes. At 1,000 feet, you see waves. It's no longer smooth; it has texture. If you splash down into the water and look at the surface with a magnifying glass, you see foam, bubbles, churning eddies, and breaking droplets. The smoothness was an illusion of distance. Down at the scale of the Planck length (the smallest possible unit of distance, 10^{-35} meters), space-time itself is not a smooth sheet. It is a foam of changing geometry, where "up" and "down" might lose their meaning, and microscopic wormholes might form and collapse in an instant.

The Active Void

This is the second great shock of the book. First, we learned that solid matter is mostly empty space. Now, we learn that empty space is full of "stuff." The vacuum is not a passive container; it is a medium that is constantly fluctuating. This "jitter" of the vacuum is the baseline rhythm of the universe. It is the heartbeat of reality.

And this is not just theory. You might be tempted to think, "Well, if these particles appear and disappear so fast that we can't see them, maybe they are just a mathematical trick." But they are not. As we will see in the next chapter, these ghostly particles can interact with the real world. They can nudge electrons, push metal plates, and perhaps even kick start the birth of the universe. The uncertainty that Heisenberg discovered doesn't just make the world fuzzy; it makes the nothingness heavy, active, and powerful. The void is alive.

Six

Particles From Absolutely Nowhere

Ghost in the Machine

By the mid-20th century, physicists had accepted the mathematical necessity of the "seething vacuum." Heisenberg's equations were clear: you could not have a perfectly quiet universe. However, science is not a discipline of faith; it is a discipline of evidence. It is one thing to say that the empty space in a box is teeming with invisible, fleeting particles that exist for a billionth of a nanosecond; it is quite another to prove it.

The challenge of detecting these "virtual particles" is inherent in their very nature. Remember the rules of the cosmic bank loan: the energy to create these particles is borrowed from nothing, and it must be paid back almost instantly. To "see" a virtual particle directly to catch it with a camera or a Geiger counter would require intercepting it before it disappears. But if you intercept it, you stop it from recombining with its partner. You prevent the repayment of the energy loan. This violates the uncertainty principle. Therefore, by definition, virtual particles cannot be observed directly. They are the perfect crime: they steal into existence, do their business, and vanish without leaving a footprint or so we thought.

While we cannot see the ghosts themselves, we discovered that we can see the furniture they knock over. Virtual particles may be invisible, but they carry electromagnetic fields, and those fields interact with the "real" matter in the universe. If you look closely enough at the behavior of real atoms, you can see them shivering in the cold wind of the quantum vacuum.

The Hydrogen Whisper

The first definitive proof came in 1947, from a physicist named Willis Lamb. He was studying the simplest object in the universe: the hydrogen atom. A hydrogen atom is just one proton and one electron. According to the best theories of the time (specifically, the Dirac equation), the electron in a hydrogen atom could occupy several different energy levels, or "orbits." The theory predicted that two specific orbits technically known as the $2S_{1/2}$ and $2P_{1/2}$ states should have exactly the same energy. They should be identical.

But when Lamb used microwave radiation to probe hydrogen atoms with extreme precision, he found something disturbing. The two states were *not* identical. One was slightly higher in energy than the other. There was a tiny, unexplained gap. In the world of physics, a tiny discrepancy is often the thread that unravels the entire sweater. This gap became known as the **Lamb Shift**.

The explanation for this shift was revolutionary. The electron in the hydrogen atom was not just orbiting the nucleus in a lonely vacuum. As it moved, it was plowing through a sea of virtual photons popping in and out of existence. These virtual particles were buffering the electron, jostling it like a small boat in a choppy sea. This constant "quantum trembling" smeared out the electron's position slightly, changing its interaction with the nucleus and shifting its energy level. Willis Lamb had not seen the vacuum fluctuations directly, but he had measured their effect. He had heard the whisper of the void.

The Art of Nothing

To make sense of this chaotic new reality, physicists needed a new language. The old equations were becoming nightmares of complexity, trying to track infinite numbers of virtual particles. Enter Richard Feynman, a bongo-playing genius who rewrote the book on how we visualize the universe.

Feynman developed a system of doodles that turned difficult calculus into simple pictures. In a **Feynman Diagram**, a straight line represents a real particle (like an electron), and a squiggly line represents a photon. But Feynman added something new: loops.

He showed that as an electron travels through space, it can briefly emit a virtual photon and then immediately reabsorb it. Or, a photon can spontaneously split into a virtual electron and a virtual positron (its antimatter twin), which loop around and crash back together into a photon. These "loops" represent the activity of the vacuum. When physicists use these diagrams to calculate the properties of particles, they have to sum up every possible loop every possible interaction with the nothingness.

When they included these vacuum loops in their calculations for the electron, the numbers matched Willis Lamb's experiments to an exquisite degree of accuracy. The theory of Quantum Electrodynamics (QED), which is built on the reality of these virtual particles, is now the most precise theory in the history of science. It predicts the magnetic properties of the electron correctly to more than ten decimal places. This is equivalent to measuring the distance from New York to Los Angeles to within the width of a single human hair. We know the vacuum is alive because the math of the vacuum is the only math that works.

Creation and Annihilation

What exactly happens in these fluctuations? It is a dance of duality. Matter and antimatter are exact opposites. When they meet, they annihilate, turning their mass entirely into energy (usually photons). The reverse is also true: if you have enough energy, you can create a matter-antimatter pair.

In the vacuum, the "bank loan" of energy creates a particle (say, an electron) and its antiparticle (a positron). They pull apart for a fleeting instant, existing as a dipole. But the debt must be paid. The attractive force between them snaps them back together. They collide and vanish, returning the energy to the vacuum. This is happening trillions of times a second in every atom of your body. You are constantly being permeated by a "ghost mist" of matter and antimatter.

Usually, this cycle is closed. Nothing gets out. But what if you could break the cycle? What if you could steal one of the partners before they crashed back together?

Cheating the Void: Hawking Radiation

In 1974, Stephen Hawking proposed a scenario where the vacuum could be tricked into making real matter. He applied quantum theory to the edge of a black hole the Event Horizon.

Imagine a virtual particle pair popping into existence right on the edge of this gravitational cliff. One particle forms just inside the horizon, and the other forms just outside. The inner particle is captured by the black hole's gravity and falls in. The outer particle, however, is on the safe side.

Usually, the pair would annihilate. But now, they are separated by an inescapable wall. They cannot recombine. The "loan" cannot be repaid in the usual way. The particle on the outside has no partner to annihilate with. It is stranded. Suddenly, it stops being a virtual particle and becomes a *real* particle. It flies away from the black hole as radiation.

But who pays the debt? The universe strictly enforces energy conservation. Since the new particle has real energy and is flying away, that energy must come from somewhere. It comes from the black hole itself. The particle that fell in effectively has "negative energy" relative to the outside universe, subtracting mass from the black hole.

This led to the shocking conclusion that black holes are not truly black. They glow with the light of the vacuum. Over billions of years, this **Hawking Radiation** will cause the black hole to evaporate completely. The vacuum is so powerful that it can eventually eat the most destructive objects in the universe. In the presence of a strong enough force (gravity), the "nothing" can become "something" permanently.

Seven

The Force of Emptiness

The Question That Shouldn't Have an Answer

In 1948, two Dutch physicists, Hendrik Casimir and Dirk Polder, were working at the Philips Research Laboratories in Eindhoven. They were investigating the properties of colloids—mixtures like milk or paint where tiny particles are suspended in fluid. Their work was practical and industrial, but the conversation took a turn into the bizarre theoretical realm of the quantum vacuum.

Casimir asked a strange question: "What would happen if you took two perfectly flat metal plates, placed them in a perfect vacuum, and moved them incredibly close together?"

According to classical physics, the answer is "absolutely nothing." If the plates are uncharged (so no static electricity) and non-magnetic, and if gravity is negligible because the plates are light, there is no known force in the universe that should act on them. They should hang there, suspended in the void, completely inert.

But Casimir, using the logic of quantum mechanics, predicted something else. He claimed that the empty space itself would grab the plates and push them together. He predicted that "nothing" would exert a physical force. This phenomenon, now known as the **Casimir Effect**, is perhaps the most elegant demonstration that the vacuum is not just a mathematical concept, but a physical reality with muscle.

The Music of the Void

To understand why the vacuum pushes, we have to stop thinking of virtual particles as little billiard balls and start thinking of them as waves. In quantum field theory, every particle is also a wave. The vacuum is full of waves of all possible sizes (wavelengths), constantly rippling in every direction.

Imagine the vacuum as a guitar string. A guitar string can vibrate with many different notes. It can have a long, deep wave that takes up the whole string, or a short, high-pitched wave that ripples quickly.

Now, consider the space *between* Casimir's two metal plates. Because the plates are conductors, they act like mirrors for electromagnetic waves. This creates a boundary condition. Just as a guitar string is tied down at both ends, limiting the waves that can fit on it, the space between the plates can only support waves that "fit" perfectly inside.

If the gap is tiny, only short, high-energy waves can fit in the gap. The long, low-energy waves are too big; they are excluded. They cannot exist in that small space.

The Pressure Difference

However, *outside* the plates, in the vast rest of the universe, there are no restrictions. Waves of all sizes—long, short, massive, tiny—are rippling freely.

This creates an imbalance.

- **Inside the gap:** There are fewer waves pounding against the plates.
- **Outside the gap:** There are more waves pounding against the plates.

Since these virtual waves carry momentum (radiation pressure), they push against anything they hit. The result is a simple pressure difference. The "pressure of nothing" is stronger on the outside than on the inside. Consequently, the plates are pushed together.

It is crucial to understand that the plates are not "attracting" each other like magnets. They are being *pushed* together by the universe. The vacuum outside is crushing them inwards because the vacuum inside is too thin to resist.

The Maritime Analogy

There is a wonderful historical analogy for this, dating back to the days of 19th-century sailing ships. Sailors noticed a mysterious phenomenon: if two large ships rocked side-by-side in a choppy sea with no wind, they would drift together and eventually collide. It was often blamed on "magnetic attraction" or superstition.

The real reason was wave physics. The rough ocean is full of waves of all sizes. But in the narrow channel of water between the two ships, the long ocean swells are blocked. The water between the ships is calmer than the water outside. The violent waves crashing against the *outer* hulls of the ships push them inward, overcoming the weaker pressure of the calm water between them.

The Casimir Effect is exactly the same, but instead of ships in an ocean of water, it is metal plates in an ocean of quantum fields.

Measuring the Ghostly Touch

For decades, Casimir's prediction remained a theory. The force he calculated was incredibly small. For two plates separated by a single micrometer (a hundredth of the width of a hair), the pressure is roughly 1.3 pascals—about the weight of a water droplet spread out over a desk. In 1948, no instrument was sensitive enough to measure it without being overwhelmed by other forces like static electricity.

It wasn't until 1997 that Steve Lamoreaux at the University of Washington finally performed the definitive experiment. Using a torsion pendulum in a high vacuum, he measured the force between a gold-plated sphere and a flat plate. The results matched Casimir's prediction to within 5%. Since then, experiments have refined the accuracy even further.

We have now measured the weight of nothingness.

The Problem with Nanomachines

Today, the Casimir Effect has moved from theoretical physics to engineering problems. We are currently building Micro-Electra-Mechanical Systems (MEMS)—tiny gears, switches, and motors the size of dust motes. At this scale, the parts are so close together that the vacuum force becomes dominant.

Engineers have to deal with "stiction." If they design a tiny machine part that gets too close to another part, the vacuum will snap them together, locking the gears permanently. The "force of nothing" is actually strong enough to break our microscopic machines. We have to design nanobots that are strong enough to fight the empty space they move through.

This discovery changed our relationship with the void forever. We proved that if you exclude the vacuum from a space, the surrounding universe will try to crush that space out of existence. Emptiness wants to be full.

Eight

The Energy You Cannot Remove

The Quest for Stillness

Throughout this book, we have been playing a game of subtraction. We started with a room and removed the furniture, then the air, then the atoms. In the last few chapters, we even tried to sweep away the invisible virtual particles, only to find that they push back. Now, we must make one final attempt to reach true "nothingness" by attacking the most fundamental property of existence: energy.

In the classical world of the 19th century, energy was synonymous with motion. Heat is simply the jittering of atoms; the hotter the object, the faster the jitter. Therefore, logic dictated that if you could cool an object down enough sucking away its heat like draining water from a bathtub you would eventually reach a point where all motion stops. This theoretical endpoint is known as **Absolute Zero** (0 Kelvin, or -273.15°C).

At Absolute Zero, the classical equations say the universe should freeze into a perfect, static sculpture. Atoms should lock into place. Electrons should halt in their tracks. The "hum" of reality should silence itself. If we could cool our empty box to this temperature, surely *then* we would finally have a true void?

The Unquiet Grave

The answer, once again, is a resounding no. Quantum mechanics denies us the peace of stillness.

The reason brings us back to Heisenberg's Uncertainty Principle. As we discussed, you cannot know a particle's position and momentum precisely at the same time. Now, imagine a particle at Absolute Zero. If it were truly stopped, we would know its momentum exactly (zero). If it were frozen in a specific location, we would know its position exactly. This would be a state of infinite precision, which is forbidden by the laws of the universe.

Therefore, nature demands that even at the coldest possible temperature, particles must retain a tiny, residual amount of motion. They must jitter, just a little bit, to keep their momentum "fuzzy." This irremovable jitter is called **Zero-Point Motion**.

We see this in the real world with Helium. Helium is the only element that remains a liquid even at Absolute Zero (at standard pressure). You cannot freeze it into a solid ice block just by cooling it. Its atoms are so light, and their zero-point jitter is so vigorous, that they refuse to settle down into a solid crystal lattice. The energy of "nothing" keeps them liquid forever.

The Quantum Spring

To understand where this energy comes from, we have to look at the favorite toy of physicists: the harmonic oscillator. Think of a weight attached to a spring, bouncing up and down.

In classical physics, you can stop the spring. You can hold the weight perfectly still at the bottom. Its energy is zero.

In quantum physics, the math changes. The energy levels of a quantum spring are like the rungs of a ladder. You can be on rung 1, rung 2, or rung 10. But unlike a normal ladder, the bottom rung is not on the ground. The lowest possible energy state the "Ground State"—hovers slightly above zero.

The formula for the energy of a quantum system is $E = (n + 1/2)\hbar$omega. If you remove all the excitation (n=0), you don't get zero. You get that stubborn 1/2. That "one-half" represents the Zero-Point Energy. It means that every quantum system, in its most relaxed, empty state, retains half a photon's worth of energy. You can never turn the switch all the way off. The dimmer switch of the universe stops at "dim," it never goes to "dark."

The Infinite Basement

This would be a minor curiosity if it only applied to single atoms. But as we learned in Chapter 4, the vacuum of space is not empty; it is filled with **fields**. And here is the kicker: mathematically, a field is treated as a collection of an infinite number of harmonic oscillators.

Imagine every point in space as a tiny spring. There is a spring for every possible frequency of light. There is a spring for radio waves, a spring for microwaves, a spring for visible light, up to X-rays and gamma rays.

- The radio-wave spring has a tiny bit of zero-point energy.
- The microwave spring has a bit more.
- The gamma-ray spring has a lot.

Now, try to calculate the total energy of empty space. You have to add up the "one-half" energy of every single frequency mode in the universe.1/2 + 1/2 +

1/2 + ...

Since there is theoretically no limit to how short a wavelength can be (until you hit the Planck length), there is an effectively infinite number of springs. When you sum up an infinite number of small energies, you get a catastrophic result: **Infinite Energy.**

The Weight of the Void

According to a naive reading of Quantum Field Theory, the air in front of your nose contains enough zero-point energy to boil all the oceans on Earth. In fact, a coffee cup of "empty" space should contain enough energy to vaporize the Milky Way.

This calculation is absurd. We know it's wrong because we are still alive. If space were that energetic, the gravitational mass of that energy (remember $E=mc^2$) would have crushed the universe into a black hole instantly.

So, for decades, physicists did something pragmatic but philosophically disturbing: they ignored it. They said, "Well, we only care about *changes* in energy. The baseline doesn't matter. Let's just redefine zero to be this infinite number and only measure what happens above it." This mathematical trick, called **Re normalization**, worked. It allowed us to build lasers and computers.

But you cannot hide from gravity. Gravity doesn't care about "changes" in energy; gravity feels the *total* weight of everything. If the vacuum has energy, gravity should see it.

The Great Crisis

This leads us to the precipice of the greatest embarrassment in the history of science. We have two great theories:

1. **Quantum Mechanics**, which says empty space should be incredibly heavy with zero-point energy.
2. **General Relativity**, which says space is incredibly light (because the universe isn't curling up into a ball).

When we compare the amount of vacuum energy predicted by our equations to the amount of vacuum energy we actually observe in the cosmos (which we will discuss in Part III), the numbers do not match.

They are not just slightly off. They are off by a factor of 10^{120}. That is a 1 followed by 120 zeros. It is the largest discrepancy between theory and observation in the history of human thought. It means that either our understanding of "nothing" is fundamentally flawed, or there is something magical canceling out that energy, leaving just a tiny whisper behind.

The vacuum is hiding a secret. It is holding a power that our math says should be infinite, but our eyes tell us is almost zero. Solving this riddle is the holy grail of 21st-century physics.

Nine

Why Reality Doesn't Collapse

The Most Embarrassing Number in Physics

In the history of science, there have been many incorrect predictions. Astronomers once thought the sun revolved around the earth; chemists once thought fire was caused by a substance called phonologist. These errors were usually corrected by better data or slightly tweaked equations. But the problem we face with the energy of empty space is not a simple rounding error. It is a catastrophe of understanding so profound that it suggests we are missing more of the picture than we are seeing.

As we established in the previous chapter, Quantum Field Theory predicts that the vacuum should be dense with zero-point energy. When we sum up all the contributions from every quantum field, the resulting number for the "weight" of empty space is approximately 10^{113} joules per cubic meter. This is a number so large it effectively means "infinite" for all physical purposes.

However, we can also look at the universe to see how much the vacuum actually weighs. We do this by observing how fast the universe is expanding. If space were heavy, gravity would slow the expansion down. Based on our observations of the cosmos (which we will detail in Part III), the actual energy density of empty space is about 10^{-9} joules per cubic meter.

Compare those two numbers:

- **Theory:** 10^{113}
- **Reality:** 10^{-9}

The difference between them is roughly 10^{120}. To visualize this error, imagine you are asked to estimate the weight of the entire visible universe every star, galaxy, black hole, and planet combined. If you guessed that the entire universe weighs as much as a single grain of sand, you would be making a smaller mistake than physicists make when they calculate the energy of the vacuum.

The Imploding Universe

Why is this a crisis? Because energy has mass. According to Einstein's General Relativity ($E=mc^2$), energy generates gravity just as matter does. If the vacuum really contained the amount of energy our quantum equations predict, the gravitational pull of empty space would be titanic.

There are two ways this could have ended, and both are fatal:

1. **The Big Crunch:** The sheer mass of the vacuum would create such intense gravity that the universe would have collapsed in on itself mere moments after the Big Bang. Space would have crushed itself into a singularity before a single atom could form.
2. **The Big Rip:** Alternatively, if the pressure of this vacuum energy acted repulsively (as dark energy does), it would have blown the universe apart so fast that electrons would have been ripped from their nuclei.

The fact that you are sitting here, reading this book on a stable planet in a stable galaxy, is proof that the vacuum energy is not huge. It is incredibly, suspiciously tiny. It is effectively zero, but and this is the maddening part *not exactly* zero.

The Cancellation Miracle

How can a calculation be so wrong? For decades, physicists pinned their hopes on a concept called **Symmetry**.

In the quantum world, particles come in two main families: **Bosons** (force carriers like photons) and **Fermions** (matter particles like electrons). When you calculate the vacuum energy, these two families contribute in opposite ways. Bosons tend to add positive energy to the vacuum, while fermions effectively add negative energy.

Physicists hoped that these two contributions might cancel each other out perfectly. Imagine a tug-of-war where one team is pulling with infinite strength to the left, and the other team is pulling with infinite strength to the right. The net result on the rope? Zero movement.

This idea blossomed into a theory called **Super symmetry (SUSY)**. Super symmetry posits that for every known particle, there is a heavy, invisible "super-partner." The electron has the "selectron"; the photon has the "photino." If this theory were true, the vacuum energy of the bosons and the fermions would perfectly balance, naturally resulting in a zero-energy vacuum. It would explain why empty space doesn't weigh anything: the "stuff" in the vacuum is perfectly canceling itself out.

The Broken Mirror

There is just one problem: We can't find them.

For the last decade, the Large Hadron Collider (LHC) at CERN has been smashing protons together, hunting for these super symmetric partners. Thousands of scientists have sifted through petabytes of data, hoping to see a glimpse of a selectron or a gluino. The result? Nothing.

We have found the Higgs Boson, but the super-partners remain missing. If Super symmetry exists, it is "broken" meaning the partners are much heavier than we thought, so they don't cancel out the vacuum energy perfectly. The beautiful cancellation fails. We are left with the massive number again, and no easy way to subtract it. The silence from the LHC has deepened the crisis. We can no longer rely on simple symmetry to clean up the mess in our equations.

The Fine-Tuning Problem

This leaves us with an uncomfortable possibility known as **Fine-Tuning**.

If there is no natural law (like Super symmetry) that automatically sets the vacuum energy to zero, then the cancellation must be happening by "accident." It implies that the positive energy and the negative energy of the vacuum are tuned to cancel each other out to within 120 decimal places of precision, leaving just a tiny sliver of energy remaining (10^{-9} joules).

Imagine throwing a dart from the other side of the galaxy and hitting a specific bullseye on Earth that is one atom wide. That is the level of precision required for the vacuum energy to be this small by chance.

Most scientists hate this explanation. It feels like "magic" or divine intervention, which science tries to avoid. It implies that the parameters of the universe were exquisitely dialed in to allow for our existence. If the vacuum energy were even slightly different if the cancellation were off by just a hair galaxies would never have formed, and life would be impossible.

The Anthropic Escape Hatch

Faced with this improbability, many physicists have turned to the **Multiverse** theory. This idea, championed by string theorists, suggests that our universe is just one bubble in a vast, foaming cosmic soup.

In this scenario, there are 10^{500} different universes. Each one has a different value for the vacuum energy.

- In Universe A, the vacuum energy is huge, and the universe collapses instantly. No life.
- In Universe B, the vacuum energy is negative, and the universe implodes. No life.
- In Universe C, the vacuum energy is massive and repulsive, ripping atoms apart. No life.

Then there is Universe D our universe. By sheer roll of the cosmic dice, this bubble ended up with a vacuum energy that is nearly, but not quite, zero. Because this condition allows stars and galaxies to form, complex chemistry can occur, and eventually, observers like us can evolve.

We look around and ask, "Why is the vacuum energy so small?" The Anthropic Principle answers: "**Because if it weren't, you wouldn't be here to ask the question.**"

It is a tautology, a circular argument, but it might be the only answer we have. We may be survivors of a cosmic lottery, living in the only habitable void in an ocean of lethal universes.

Ten

The Biggest Discovery Since Gravity

The Fate of the Universe

By the late 1990s, cosmology had settled into a comfortable, albeit slightly depressing, consensus. We knew the universe began with the Big Bang approximately 14 billion years ago. We knew it had been expanding ever since, flinging galaxies apart like shrapnel from a grenade. The great unanswered question was not *what* was happening, but *how it would end.*

The logic was simple, based entirely on Isaac Newton's law of gravity. Gravity is an attractive force; it pulls things together. Therefore, the mutual gravitational attraction of all the stars, galaxies, and dark matter in the universe must be acting as a cosmic brake. Just as a ball thrown into the air eventually slows down because Earth's gravity pulls it back, the expansion of the universe should be slowing down as the galaxies pull on one another.

The destiny of the cosmos, cosmologists believed, depended on a single variable: density. If there was enough "stuff" (mass) in the universe, gravity would eventually win the tug-of-war, halting the expansion and pulling everything back into a "Big Crunch" a fiery reverse of the Big Bang. If there wasn't enough mass, the universe would expand forever, but it would still slow down, coasting gently into a cold, dark eternity. The mission of 1990s astronomy was to measure this deceleration. They wanted to know how hard the brake pedal was being pressed. Nobody absolutely nobody expected to find that someone was pressing the gas.

The Standard Candle

To measure the speed of the universe's expansion over time, astronomers needed a way to look back into the past. Because light takes time to travel, looking at distant galaxies is effectively looking back in time. If we could measure how fast a galaxy was moving away from us billions of years ago and compare it to how fast galaxies are moving today, we could calculate the deceleration rate.

But measuring distance in space is notoriously difficult. A bright star might be close, or it might be a massive star very far away. To solve this, astronomers use "Standard Candles" cosmic objects of known brightness. The specific candle chosen for this high-stakes measurement was the **Type Ia Supernova**.

A Type Ia Supernova occurs when a white dwarf star—the dense corpse of a sun-like star steals too much matter from a companion star. When the white dwarf reaches a precise critical mass (about 1.4 times the mass of our sun), it becomes unstable and detonates. Because the explosion always happens at the exact same mass limit, the brightness of the explosion is remarkably consistent. It is a cosmic light bulb of known wattage. If you see a Type Ia Supernova that looks dim, you know it is far away. If it looks bright, you know it is close. By comparing how bright these explosions appeared to how much their light had been stretched (red shifted) by the expansion of space, astronomers could map the history of the universe's growth.

The 1998 Shock

Two rival teams of astronomers set out to do this measurement: the Supernova Cosmology Project, led by Saul Perl mutter, and the High-Z Supernova Search Team, led by Brian Schmidt and Adam Riess. They spent years scanning the darkest patches of the sky, hunting for these rare explosions in distant galaxies. It was grueling work, requiring precise coordination of telescopes around the world to catch the supernovae before they faded.

By 1998, both teams had gathered enough data to plot the curve of cosmic history. They fed the numbers into their computers, expecting to see a curve that bent downward, indicating a slowing universe. Instead, the data points curved upward.

The distant supernovae were dimmer than they should have been. This meant they were farther away than the "slowing down" theory predicted. The universe had not expanded less than expected; it had expanded *more*. The expansion was not slowing down under the influence of gravity; it was speeding up.

The researchers were terrified. A result this contradictory to established physics usually means you made a mistake a dirty lens, a bad calculation, or a software bug. They checked and re-checked their work, desperately looking for the error. They found none. Reluctantly, they published their findings. The conclusion was inescapable: about 5 billion years ago, the universe shifted gears. Gravity lost its dominance, and a mysterious, invisible pressure began pushing the galaxies apart faster and faster.

The Return of Lambda

The discovery sent the physics world scrambling for an explanation. What could possibly be strong enough to overpower gravity on a cosmic scale? The answer brought them back to the "empty" space we discussed in Part I.

Decades earlier, Albert Einstein had included a term in his General Relativity equations called the **Cosmological Constant** (denoted by the Greek letter Lambda, Lambda). At the time, Einstein thought the universe was static (neither expanding nor contracting). To make his equations work, he needed a force to counteract gravity, so he invented Lambda a repulsive energy inherent to space itself that pushed outward. When the expansion of the universe was discovered in the 1920s, Einstein abandoned Lambda, calling it his "greatest blunder."

But the 1998 data resurrected the blunder. It turned out Einstein was right, but for the wrong reason. There *is* an energy inherent to empty space. As the universe expands, it creates more space. Since this energy is a property of space, more space means more repulsive energy. It is a runaway feedback loop. Gravity gets weaker as galaxies move apart, but the vacuum energy stays constant per unit of volume, meaning its *total* power grows.

Dark Energy

We gave this mysterious force a name: **Dark Energy**. It is a placeholder name for our ignorance, similar to "Dark Matter." We don't know what it is, but we know what it does. It is the physical manifestation of the nothingness.

This discovery fundamentally changed our inventory of reality. Before 1998, we thought the universe was made of atoms (stars, gas, planets) and maybe some Dark Matter. After 1998, the pie chart of the universe was rewritten.

- **Ordinary Matter (Atoms):** ~5%
- **Dark Matter:** ~27%
- **Dark Energy:** ~68%

Let that sink in. Everything you have ever seen, touched, or loved every star, every planet, every book, every person makes up only 5% of the universe. The rest is darkness. And the vast majority, 68%, is the energy of empty space itself. The "nothing" is not just a minor background detail; it is the dominant substance of the cosmos. The universe is mostly void, and that void is in control.

The Lonely Future

The implications of an accelerating universe are profound and somewhat terrifying. If the expansion continues to speed up, the sky will eventually empty out. Billions of years from now, distant galaxies will be pushed away from us faster than the speed of light. They will cross the "cosmic horizon" and vanish from view forever.

Future civilizations, if they arise on a planet in our galaxy 100 billion years from now, will look up and see. nothing. The other galaxies will be gone. The evidence of the Big Bang will have been stretched into unpredictability. They will believe they are alone in a static, island universe, with no way of knowing the rich, violent history that came before.

The discovery of Dark Energy confirmed that the vacuum is the most powerful force in existence. It has the power to defeat gravity. It has the power to isolate galaxies. And ultimately, it has the power to dictate the final death of the universe.

Eleven

Dark Energy—The Weight of the Void

The Invisible Elephant in the Room

If you were to take a survey of the universe, counting every proton, neutron, and electron that makes up the stars, nebulae, planets, and people, you would come to a humbling conclusion: you have missed almost everything. As we touched upon in the previous chapter, "normal" matter the stuff of the Periodic Table accounts for a mere 5% of the total energy budget of the cosmos. Dark Matter, the invisible scaffolding that holds galaxies together, makes up about 27%. But the remaining 68% the vast majority of reality is composed of Dark Energy.

This means that Dark Energy is not some rare, exotic substance found only at the edges of black holes or in the early moments of the Big Bang. It is the most common thing in existence. It is everywhere. It is in the space between galaxies, but it is also in the space between your eyes and this page. Inside the volume of your own body, there is a tiny amount of Dark Energy. It is the ocean in which we all swim, yet we are completely oblivious to it because it interacts with nothing. It does not reflect light, it does not cluster into clumps, and it does not react to chemistry. It simply *is*. It is the smooth, uniform background hum of the universe, and its only defining characteristic is that it refuses to be diluted.

The Weight of a Shadow

How much does this invisible ocean weigh? We can calculate its density based on how fast it is pushing the universe apart. The answer is incredibly small: approximately 6 times 10^{-10} joules per cubic meter.

To put that in perspective, this is roughly equivalent to the mass of a few hydrogen atoms spread out over the volume of the Earth. It is a whisper of energy so faint that, on the scale of a solar system or a galaxy, it is completely negligible. Gravity easily overpowers it. That is why our solar system isn't expanding; the gravity of the Sun holds the planets in a tight grip, overwhelming the tiny push of the Dark Energy inside the solar system.

However, the power of Dark Energy lies in its ubiquity. While gravity is concentrated around clumps of matter (like stars and planets), Dark Energy is everywhere. When you sum up that tiny 10^{-10} joules over the trillions of cubic light-years of empty space between the galaxies, the total amount becomes staggering. It outweighs all the stars and galaxies combined. It is a classic case of quantity over quality. The void is weak, but there is so much of it that it dictates the gravitational structure of the entire universe.

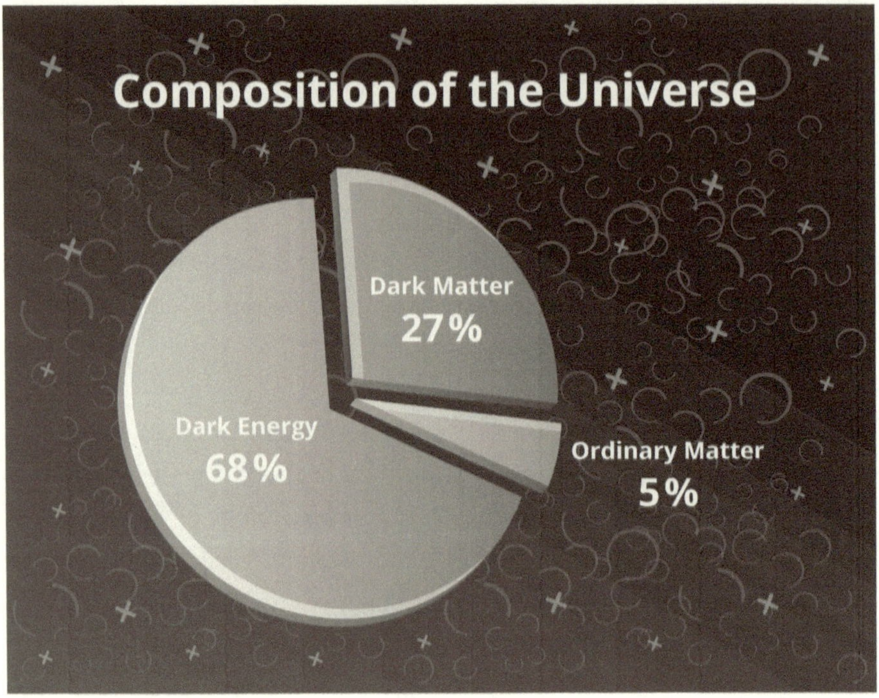

Suspect No. 1: The Cosmological Constant

So, what is it? The leading theory is the one Einstein accidentally stumbled upon: the **Cosmological Constant** (represented by Lambda or "Lambda").

In this view, Dark Energy is not a "stuff" or a fluid; it is an intrinsic property of space itself. It is the "cost" of having space. Just as water is wet, space is expansive. The defining feature of the Cosmological Constant is that its density never changes.

- If you take a box of gas and expand it to double its size, the gas gets half as dense. The particles are spread thinner.
- If you take a box of Dark Energy and expand it to double its size, the density stays exactly the same.

This seems impossible where does the new energy come from? It comes from the work done to expand the space. As the universe grows, it creates more space, and since every unit of space comes with a built-in energy cost, the total amount of Dark Energy grows with it. This creates the runaway effect: more space —> more repulsive energy —> faster expansion —> even more space. If this theory is correct, the universe will expand exponentially forever, leading to a cold, lonely, eternal night.

Suspect No. 2: Quintessence

Not everyone is happy with the Cosmological Constant. Its static nature feels too rigid for some physicists. The alternative theory is called **Quintessence**, named after the classical "fifth element" of ancient philosophy.

In this model, Dark Energy is not a built-in property of space, but a dynamic quantum field that permeates the universe, much like the electromagnetic field or the Higgs field. The key difference is that Quintessence can change over time. It can evolve.

Imagine a ball rolling down a very shallow hill. As the ball (the field) rolls, its potential energy drives the expansion of the universe. But unlike a constant, this field could eventually reach the bottom of the hill and stop. Or, it could change its character entirely. Some versions of Quintessence theories suggest that Dark Energy could eventually switch signs turning from a repulsive force back into an attractive one. If this happens, the accelerating expansion could stop and reverse, saving us from the eternal dark and leading to a "Big Crunch" after all. However, current data suggests the field is behaving suspiciously like a constant, giving us little hope for this reprieve.

Suspect No. 3: We Are Wrong About Gravity

There is a third, more radical possibility: Dark Energy doesn't exist at all.

What if our observations of accelerating expansion are not caused by a mysterious invisible energy, but by a failure in our understanding of gravity? Einstein's General Relativity has been tested continuously within our solar system, and it works perfectly. But we have never tested it on the scale of billions of light-years.

It is possible that gravity behaves differently over massive distances. This approach is called **Modified Gravity**. Theorists tweak Einstein's equations to see if they can create a universe that speeds up without adding invisible energy. While mathematically interesting, these theories often struggle to explain other things, like the Cosmic Microwave Background radiation or the way galaxies cluster. For now, General Relativity remains the king, and Dark Energy remains the necessary ghost in the machine.

The Lambda-CDM Model

Despite not knowing what Dark Energy is, we have built a remarkably successful model of the universe around it. It is called the **Lambda-CDM Model**.

- **Lambda (Lambda):** Stands for Dark Energy (the Cosmological Constant).
- **CDM:** Stands for Cold Dark Matter.

This is the "Standard Model" of the Big Bang. It is a strange triumph of science. We have a model that predicts the age of the universe (13.8 billion years), the distribution of galaxies, and the patterns of radiation left over from the beginning of time with incredible precision. And yet, the two main components of the model Lambda and CDM are complete mysteries to us.

We are like mapmakers who have perfectly charted the coastline of a continent we have never set foot on. We know the shape of the universe, its expansion rate, and its destiny, but we do not know what it is made of. We are measuring the weight of a ghost.

The Coincidence Problem

There is one final puzzle that keeps cosmologists awake at night. Why is the amount of Dark Energy and Dark Matter roughly comparable *right now*?

- Early in the universe, matter was dense, and Dark Energy was negligible.
- In the far future, Dark Energy will be dominant, and matter will be negligible.

But right now, in this brief slice of cosmic time when humans happen to exist, the two are within the same order of magnitude (68% vs 27%). This is called the **Cosmic Coincidence Problem**. Are we simply lucky to live in the "Goldilocks era" where the universe is balanced between gravity and expansion? Or is there a deeper connection between the void and the matter that we are missing? Some physicists believe this coincidence suggests that Dark Energy and Dark Matter might be two sides of the same coin a unified "Dark Fluid" that we have yet to comprehend.

Twelve

How Nothing Pushes

The Paradox of Pressure

In our everyday experience, pressure is a pushing force. If you blow up a balloon, the air pressure inside pushes outward, stretching the rubber skin. If you dive to the bottom of a swimming pool, the water pressure pushes inward against your eardrums. In both cases, pressure is positive; it exerts a force from a region of high density to a region of low density (or against a container).

So, when cosmologists say that Dark Energy accelerates the expansion of the universe because it has **negative pressure**, it sounds like nonsense. How can pressure be negative? And even if it were negative, shouldn't negative pressure imply a *suction* force? If you suck the air out of a straw, the negative pressure pulls the liquid up. Logic dictates that if the universe is filled with negative pressure, the galaxies should be sucked together, not blown apart.

To understand why the universe does exactly the opposite, we have to leave our intuition at the door and step into the equations of Albert Einstein. In

General Relativity, gravity is not just caused by mass. It is caused by three things: mass, energy, and pressure.

Gravity's Hidden Source

We are taught in school that mass creates gravity. The Earth pulls on the Moon because the Earth is massive. But Einstein showed that mass and energy are equivalent ($E=mc^2$). Therefore, energy also creates gravity. A hot brick weighs slightly more than a cold brick because the thermal energy adds to its gravitational field.

But the third component pressure is the game changer. In Einstein's field equations, pressure contributes to the gravitational field just like mass density does.

- **Positive Pressure** (like a compressed gas) creates **attractive** gravity. This is weird but true: if you have a box of gas under extreme pressure, its gravity is stronger than a box of cold gas with the same mass. The pressure pulls things inward.
- **Negative Pressure** (tension) creates **repulsive** gravity.

This is the key. Dark Energy has negative pressure. It is not like a gas pushing outward; it is like a stretched rubber sheet trying to snap back. But because this "tension" exists everywhere in the universe, its gravitational effect is repulsive. It creates a "push" that counteracts the attractive gravity of matter.

The Suction That Blows

Let's try to visualize this with a piston. Imagine a cylinder filled with normal gas. If you pull the piston out, the gas pressure drops. The gas wants to expand, but its density decreases as you increase the volume.

Now, imagine a cylinder filled with Dark Energy (vacuum energy). As we learned, the density of vacuum energy is constant. If you pull the piston out to double the volume, you have twice as much vacuum energy as you started with. You have *created* energy by increasing the volume.

Thermodynamics tells us that energy cannot come for free. If the energy inside the cylinder goes up, you must have done "work" on the system. You had to pull hard on the piston to expand it. This resistance you feel the vacuum trying to pull the piston back in is the **negative pressure**.

So, mechanically, the vacuum behaves like a suction; it wants to collapse the piston. But gravitationally, this tension has the opposite effect. In General Relativity, this massive amount of negative pressure generates a gravitational field that pushes everything apart.

It is a double negative:

1. The vacuum tries to pull itself together (tension).
2. Gravity sees this tension and flips it into a repulsive force (anti-gravity).

The result is that the "suction" of the vacuum actually drives the galaxies away from each other.

The Runaway Train

This mechanism creates a terrifying feedback loop.

In a normal explosion (like a stick of dynamite), the force is strongest at the beginning and then fades away as the debris scatters. The pressure drops as the cloud expands.

The Big Bang was not like dynamite. Because the expansion is driven by empty space, and expansion *creates* more empty space, the engine of expansion never runs out of fuel.

- The universe expands.
- This creates more volume of vacuum.
- More vacuum means more Dark Energy.
- More Dark Energy means more repulsive force.
- The universe expands faster.

This leads to exponential growth. We are currently in the era where Dark Energy has taken the wheel. For the first 7 billion years of the universe, gravity was winning; the expansion was slowing down. But as the matter spread out and diluted, the constant density of Dark Energy eventually overtook it. We passed the "tipping point" about 5 billion years ago. We are now on a runaway train that is picking up speed with every passing second.

The Phantom Menace

It could be worse. The standard theory (Cosmological Constant) says the density of Dark Energy is constant ($w = -1$). This leads to a "Big Freeze," where galaxies simply get too far apart to see each other.

But there is a theoretical possibility called Phantom Energy where the density of Dark Energy actually increases over time ($w < -1$).

If this is true, the repulsive force won't just push galaxies apart. It will eventually become strong enough to overcome the gravity holding the Milky

Way together. The galaxy will scatter.

Then, it will overcome the gravity of the solar system. The Earth will be ripped from the Sun.

Then, it will overcome the electromagnetic forces holding atoms together.

finally, it will rip the nuclei of atoms apart.

This scenario is called the **Big Rip**. In the final fraction of a second, the fabric of spacetime itself would be torn asunder by the sheer violence of the nothingness. While current data suggests we are safe from the Big Rip (Dark Energy seems stable), the fact that the equations allow for it is a reminder of how little we understand the monster we have discovered.

The Ultimate Irony

There is a profound irony in all of this. For thousands of years, philosophers and theologians argued that "from nothing, nothing comes" (*ex nihilo nihil fit*). Science has proven this wrong in the most dramatic way possible.

Not only can "something" come from nothing (virtual particles), but the "nothing" itself is the most energetic, dynamic, and powerful driver of cosmic evolution. The void is not the passive absence of creation; it is the engine of the future. We are being pushed into the dark by the very emptiness that surrounds us.

Thirteen

Before the Beginning

The Uniformity Mystery

If you point a sensitive radio telescope at the night sky, you can detect the faint afterglow of the Big Bang the Cosmic Microwave Background (CMB). This radiation is a snapshot of the universe when it was just 380,000 years old, a time when the hot plasma of creation finally cooled enough to let light travel freely. When astronomers first mapped this baby picture of the cosmos, they found something deeply unsettling: it was too perfect.

No matter where they looked left, right, up, down the temperature of this background radiation was almost exactly the same: 2.725 Kelvin. It varied by no more than one part in 100,000. This uniformity is known as the **Horizon Problem**, and it kept cosmologists awake at night for decades.

Here is why it is a problem: Imagine a cup of hot coffee and a cup of cold milk sitting on a table. If you pour them together, they eventually mix and reach a uniform warm temperature. But this mixing takes time. The heat has to travel from the coffee to the milk. In the early universe, regions of space on opposite sides of the sky were separated by vast distances—distances so great that light had not had enough time to travel between them since the Big Bang began. There was no way for "Point A" to send a signal to "Point B" to say, "Hey, I'm at 2.725 Kelvin, you should be too." They were causally disconnected. Yet, somehow, they were at exactly the same temperature.

It was as if you walked into a party and found 100 people who had never met, never spoken, and lived on different continents, all wearing the exact same outfit down to the matching socks. It wasn't a coincidence; it was a clue that we were missing a massive piece of the story.

The Blow-Up

In 1979, a physicist named Alan Guth proposed a radical solution. He suggested that the Big Bang we know—the hot expansion of matter—was actually preceded by a brief, violent burst of "super-expansion" called **Cosmic Inflation**.

Guth theorized that for a fleeting fraction of a second (roughly from 10^{-36} to 10^{-32} seconds), the universe did not just expand; it ballooned. It doubled in size, then doubled again, and again, doubling perhaps 100 times in a blink. In less time than it takes light to cross the nucleus of an atom, the universe grew from the size of a proton to the size of a grapefruit (and likely much, much larger).

This solved the Horizon Problem instantly. It meant that the entire observable universe everything we see today started as a microscopic speck that *was* connected. The coffee and milk had mixed when the cup was tiny. Then, Inflation took that tiny, uniform speck and stretched it out faster than light, blowing it up to cosmic proportions. The uniformity we see in the sky today is simply the stretched-out fingerprint of that original microscopic contact.

The Inflation Field

What drove this insane expansion? It wasn't heat, and it wasn't normal matter. It was, once again, the vacuum.

Guth proposed a new quantum field called the **Inflation Field**. In the very first moments of time, this field was stuck in a state of high energy a "False Vacuum." It wanted to decay to a lower energy state (a "True Vacuum"), but it was momentarily trapped, like a ball stuck in a dip on a high mountain shelf.

While it was stuck there, the Inflation Field filled space with a massive amount of potential energy. As we learned in the previous chapter, energy embedded in space creates **negative pressure**. But unlike the gentle Dark Energy of today, this primordial vacuum energy was titanic. It created a repulsive gravity so powerful that it tore space apart exponentially.

This was the "Bang" in the Big Bang. It wasn't an explosion of matter into space; it was an explosion of space itself, driven by the pent-up energy of a vacuum that wanted to settle down.

The Ultimate Free Lunch

But where did all the "stuff" come from? If the universe started as empty space expanding rapidly, how did we end up with stars, galaxies, and gas?

This brings us to the most magical moment in physics: **Reheating**. Eventually, the Inflation Field "rolled down" the hill. It decayed from its high-energy state to its low-energy state. When a quantum field drops in energy, that energy has to go somewhere. It cannot just disappear.

The energy that was driving the expansion was suddenly dumped into the other quantum fields the quark fields, the electron fields, the photon fields. The vacuum energy was converted into particles. The universe, which had been cold and empty during inflation, suddenly became a hot, dense soup of matter and radiation. The expansion slowed down to a normal, coasting speed, and the hot Big Bang began.

This implies something staggering: All the matter in the universe every atom in your body, every star in the sky is just congealed vacuum energy. We are the leftovers of a quantum field that collapsed 13.8 billion years ago.

Alan Guth famously said, "**The universe is the ultimate free lunch.**" He meant that the positive energy of matter (mass) is exactly balanced by the negative energy of gravity.

- Mass/Energy is positive (+1).
- Gravity is attractive, so gravitational potential energy is negative (-1).
- Total Energy of the Universe = (+1) + (-1) = 0.

The universe did not require a massive injection of energy to be born. It required zero net energy. It could emerge from nothing because, thermodynamic ally speaking, it is still nothing. It is just "nothing" that has been separated into positive and negative parts.

The Quantum Seed

But there is one final detail that brings the story full circle. If Inflation smoothed everything out, why are there galaxies at all? Why isn't the universe just a perfectly smooth, boring soup of gas?The answer lies in the uncertainty principle. During inflation, the Inflation Field was subject to quantum fluctuations. Tiny jitters. Microscopic ripples in the energy density.

Normally, these ripples would appear and vanish like the virtual particles we discussed earlier. But the expansion of space was so fast that it caught these ripples and stretched them. It took a quantum fluctuation smaller than an atom and stretched it until it was light-years across.

The tiny "bumps" in the quantum vacuum were frozen into the sky. When inflation ended, these bumps became regions of slightly higher density. Over billions of years, gravity pulled more matter into these denser clumps, eventually forming galaxies and stars.

When you look at the map of the Cosmic Microwave Background, those blue and red splotches are not just random noise. They are photographs of quantum fluctuations from the beginning of time. You are looking at the quantum vacuum, magnified by a factor of 10^{30}. We are here because the nothingness jittered, and inflation refused to let it smooth back out.

Fourteen

The Quantum Origin

The Problem of "Before"

We have now traced the history of the universe back 13.8 billion years, past the formation of stars, past the formation of atoms, all the way to the violent flash of Inflation that occurred at 10^{-35} seconds. But human curiosity is insatiable. We inevitably ask the question that drives physicists crazy: "What happened *before* that?"

In the classical model of the Big Bang, this question is considered meaningless. If time was created along with space at the moment of the singularity, then there was no "before." Asking what came before the Big Bang is like asking what is north of the North Pole. You can walk north as long as you like, but once you reach the pole, the concept of "north" ceases to exist. There is nothing north of the North Pole not because it is a void, but because the geometry of the Earth makes the question invalid.

However, quantum mechanics hates singularities. It refuses to accept a point of infinite density where the laws of physics break down. Therefore, many modern physicists believe that there was not a sharp "beginning" at all, but rather a quantum transition. To understand the origin of the universe, we must stop thinking of the Big Bang as an explosion *in* time, and start thinking of it as the nucleation *of* time.

Tunneling from Nowhere

In the quantum world, particles can do impossible things. If you roll a ball up a hill and it doesn't have enough speed to reach the top, it will roll back down. Classical physics forbids it from crossing the hill. But in quantum mechanics, there is a small probability that the ball will simply disappear from one side of the hill and reappear on the other side, having passed through the solid barrier. This is called **Quantum Tunneling**.

In the 1980s, the physicist Alexander Vilenkin proposed a mind-bending application of this principle. He suggested that the universe itself might have "tunneled" into existence.

Imagine a state of absolute nothingness—no space, no time, no matter. In quantum theory, even "nothing" has a wave function. Vilenkin showed that the laws of quantum mechanics allow for a "Universe from Nothing" event. Just as a radioactive atom can spontaneously decay without any external trigger, the void can spontaneously "decay" into a small, closed universe.

In this model, the universe didn't start as a singularity. It started as a tiny bubble of spacetime that popped into existence out of the quantum potential. It didn't need a cause. It appeared because the laws of physics make it probable for it to appear given enough "time" (though "time" here is a metaphor). The universe exists simply because zero is unstable.

The No-Boundary Proposal

Stephen Hawking and James Harte took a different, perhaps even more elegant approach. They wanted to eliminate the messy singularity entirely. They proposed the **No-Boundary Condition**.

To explain this, they used the concept of **Imaginary Time**. This sounds like science fiction, but it is a standard mathematical tool. Real time is what we experience—a line moving from past to future. Imaginary time runs perpendicular to real time. When you look at the universe through the lens of imaginary time, the distinction between time and space disappears. Time becomes just another dimension of space.

In the Harte-Hawking model, if you travel backward in time toward the Big Bang, you don't hit a sharp wall or a singularity. Instead, time rounds off, like the bottom of a shuttlecock or the tip of a sphere. Just as the dimensions of a sphere get smaller as you move toward the pole but never "break," time simply becomes space.

This means the universe is finite but has no boundary. It has no beginning edge. It is a self-contained, 4-dimensional shape. In this view, the question "What created the universe?" is replaced by the realization that the universe just *is*. It doesn't need a creator or a trigger because it doesn't have a start point in the traditional sense. It is a closed loop of existence.

The Probability of You

If the universe emerged from a quantum process, then it is subject to the laws of probability. Quantum mechanics doesn't predict exactly what will happen; it predicts a range of possible outcomes.

This implies that the "Quantum Origin" didn't just create *one* universe. It likely created a wave function containing infinite possible universes.

- In some, the laws of physics are different.
- In some, gravity is repulsive.
- In some, electrons weigh as much as bowling balls.

We are living in the "most probable" history that allows for observers, or perhaps we are just one outcome in a vast branching tree of realities. This connects back to the Many-Worlds Interpretation of quantum mechanics. The origin of the universe might not be a single unique event, but a continuous probability distribution.

Something Rather Than Nothing

Why does this matter? Because it shifts the burden of existence from a "Creator" to a "Law."

In the religious or classical view, you need a First Mover to set the ball rolling.

In the Quantum Origin view, the laws of physics themselves are the First Mover.

But this leads to the deepest philosophical question of all, one that physics cannot answer: Where did the laws come from?

Even if the universe popped into existence via a quantum fluctuation, that fluctuation followed the rules of quantum mechanics. Who wrote those rules? Does the "law of gravity" exist outside of the universe? Did the potential for existence exist before existence itself?

We have peeled back the layers of the onion. We found atoms, then protons, then quarks, then fields, then the quantum vacuum. Now we have reached the mathematical logic that governs the vacuum. We have explained *how* nothing became something, but we are still left staring at the mystery of *why* the logic exists at all.

Fifteen

The Vacuum That Could End Us

The False Floor

Throughout this book, we have treated the vacuum as the bedrock of reality. We have assumed that the "empty space" we inhabit the state where fields are at their lowest possible energy is the absolute bottom. We assume that if you take all the energy out of a system, you hit the ground floor and can go no lower.

But what if we are wrong? What if we are not on the ground floor, but on a mezzanine?

In physics, a system is "stable" if it is in its lowest possible energy state. Think of a ball-sitting at the very bottom of a valley. If you kick it, it rolls back down. It is secure.

However, a system can also be "meta stable." Imagine a ball sitting in a small dip halfway up a mountain. It looks stable. It can sit there for a million years. But if you give it a hard enough shove, or if it tunnels through the ridge holding it back, it will plunge down to the real valley floor far below.

This terrifying possibility applies to the universe itself. The properties of our vacuum specifically the **Higgs Field**, which gives particles their mass might be stuck in a "False Vacuum." We might be living in that precarious dip on the mountainside, blissfully unaware that a deeper, "True Vacuum" exists below us. If the universe ever discovers a way to fall into that lower state, our reality will end instantly.

The Bubble of Doom

How would this catastrophe happen? It wouldn't require a cosmic collision or a black hole. It would happen via **Quantum Tunneling**.

According to the Standard Model of particle physics, there is a non-zero probability that at any moment, in any random point in the cosmos, a tiny patch of the Higgs field could spontaneously tunnel through the energy barrier and drop to the True Vacuum state.

This would create a "bubble" of the new reality. This is not a bubble of matter; it is a bubble of different physics.

Inside this bubble, the laws of nature would be rewritten. The value of the Higgs field would be different, meaning the masses of all elementary particles would change.

- Electrons might become heavy, causing all atoms to collapse.
- Or they might become mass less, causing atoms to fly apart.
- The constants of nature, like the speed of light or the strength of electromagnetism, could shift.
- Chemistry would become impossible.

This event is known as **Vacuum Decay**. The bubble would not stay small.

Driven by the enormous release of energy from the field dropping to a lower state, the bubble would expand. In fact, it would expand at the speed of light.

The Silent killer

The speed of the expansion is what makes Vacuum Decay uniquely terrifying. Because the wall of the bubble moves at the speed of light, no information can travel faster than it. You can never see it coming.

If a Vacuum Decay bubble formed in the Andromeda Galaxy right now, we would not know. We would look up and see Andromeda shining as usual. The light from the galaxy and the wall of the destruction would travel toward us side-by-side.

Millions of years later, at the exact moment the light from the event reached Earth, the wall would hit us. There would be no warning. No looking up at the sky and seeing a growing dark spot. No time to say goodbye. One millisecond, we are here, made of stable atoms, thinking and feeling. The next nanosecond, the laws of physics that hold our bodies together would evaporate. We would not even feel pain, because the nerve impulses that carry pain signals rely on chemistry that would no longer exist. We would simply cease to be. The universe would be overwritten.

Living on the Edge

Is this just a ghost story, or is it real science? For a long time, we didn't know. The stability of the vacuum depends on the specific masses of two particles: the **Higgs Boson** and the **Top Quark**.

If the Higgs is heavy and the Top Quark is light, the vacuum is stable. We are safe.

If the Higgs is light and the Top Quark is heavy, the vacuum is unstable.

If they are in the middle, we are "meta stable"—balancing on the edge of the cliff.

When we finally measured the mass of the Higgs Boson at the Large Hadron Collider in 2012 (approx. 125 GeV) and combined it with the mass of the Top Quark (approx. 173 GeV), the results were unsettling. We landed right in the "meta stable" zone.

The data suggests that our universe is indeed living on borrowed time. The vacuum is not the absolute bottom. It is likely that in the extremely distant future perhaps 10^{100} years from now the universe will decay. However, because it is a quantum process, it *could* theoretically happen tomorrow. The probability is infinitesimally low, but it is not zero.

The Ultimate Fragility

This concept brings our exploration of "nothingness" to its ultimate climax. We began this book by thinking of empty space as a harmless void. We learned it is a soup of particles, a source of gravity, and the origin of the Big Bang. Now, we realize it is also a loaded gun pointed at the head of reality.

Our existence relies entirely on the nothingness maintaining its current, precarious state. We are like ice skaters on a frozen lake, unaware that the ice is actually a super cooled liquid waiting for a trigger to freeze solid.

But there is a strange comfort in this. If Vacuum Decay happens, it destroys everything not just us, but the atoms, the stars, and the galaxies. It cleans the slate. It replaces our universe with a new region of space with different laws, perhaps laws that allow for new, stranger forms of complexity to arise. We are temporary patterns in a vacuum that is restless, constantly seeking a lower, simpler state of being.

Sixteen

The Hardest Question

The Child's Nightmare

There is a moment in the cognitive development of almost every human being, usually around the age of six or seven, when the infinite crashes into the finite. It happens late at night. The child lies in bed, thinking about the world, and asks, "Where did God come from?" or "What was there before the universe?" When the parents offer the standard answer "Nothing" the child tries to simulate this in their mind. They try to imagine a state of total absence.

They might picture a dark room, or a blank sheet of paper, or a sleeping dreamless sleep. But the mind recoils. Even a dark room has space. Even a blank paper has color. Even sleep implies a sleeper. The human brain, evolved to process sensory input, is incapable of running a simulation of non-existence. We encounter a cognitive error message. We feel a sudden, vertiginous drop in our stomach a sense of cosmic horror.

This specific type of dread arises because we have stumbled upon the limit of human reason. We have asked the question that has haunted every civilization since the dawn of language. It is the question that drove the invention of gods, the writing of genesis myths, and the building of particle accelerators.

Leibniz's Challenge

In 1697, the German polymath Gottfried Wilhelm Leibniz, one of the inventors of calculus, crystallized this anxiety into a single, razor-sharp sentence. He didn't ask how the planets moved or what matter was made of. He asked:

"Why is there something rather than nothing?"

Leibniz argued based on the Principle of Sufficient Reason the idea that for every fact, there must be a reason why it is so and not otherwise. If the universe exists, there must be a reason for its existence.

If the natural state of things is "nothingness" because nothingness is simpler, easier, and requires no explanation then the existence of "something" is a deviation that demands a cause.

Why does the universe go to the trouble of existing? Why are there stars? Why are there laws of physics? Why is there a reader reading this sentence? The burden of proof, Leibniz argued, is on the *something*. The universe is an imposition. It is a massive, complex, energetic violation of the silence.

The Physicist's Answer (and Why it Fails)

For centuries, theology answered this question ("God made it"). But in the last few decades, physicists have stepped into the pulpit. Based on the science we have covered in the previous fifteen chapters, modern physics offers a confident rebuttal to Leibniz.

The physicist says: "We have solved it. We know why there is something. There is something because the laws of quantum mechanics dictate that 'nothing' is unstable."

They point to the quantum vacuum. They explain how virtual particles pop into existence. They explain how the Inflation field can blow a microscopic bubble of space into a galaxy-filled cosmos. They argue that the total energy of the universe is zero, so no "miracle" is needed to create it. To a physicist like Lawrence Krauss (author of A Universe from Nothing), the question is obsolete. The universe exists because the vacuum naturally spits out universes.

But to the philosopher, this answer is a cheat. It is a bait-and-switch.

The Two Nothings

The conflict arises because physicists and philosophers are using the same word to mean two completely different things.

1. The Physicist's Nothing:
To a scientist, "nothing" is a physical state. It is a quantum vacuum. It is empty space. It is a set of fields with the value of zero. It is governed by laws (quantum mechanics, relativity, gravity). As we have seen, this "nothing" is incredibly complex. It has structure, potential, and energy. It is definitely something.

2. The Philosopher's Nothing:
To a philosopher, "nothing" means non-being. It means the absence of everything including the laws of physics. It means no space, no time, no quantum fields, no potential, no math, no logic. When a physicist says, "The universe came from nothing because quantum laws allow it," the philosopher asks, "But where did the quantum laws come from?" Laws are something. Mathematical relationships are something. If you start with a quantum

field, you haven't started with nothing; you've started with a per-existing structure that governs reality. You have just kicked the can down the road. You explained the origin of matter, but you haven't explained the origin of the rules that create matter.

The Infinite Regress

This leads us into a trap called "Infinite Regress."

- **Child:** Why is there a universe?
- **Parent:** Because of the Big Bang.
- **Child:** What caused the Big Bang?
- **Parent:** A quantum fluctuation in the vacuum.
- **Child:** Where did the vacuum come from?
- **Parent:** The laws of Quantum Field Theory.
- **Child:** Where are the laws written? Who enforces them? Why *those* laws and not different ones?

Eventually, you hit a wall. You must arrive at something that has no cause—something that just is.

In religion, this uncased cause is God.

In science, this uncased cause is the Laws of Physics (or the Multiverse).

But neither answer truly satisfies Leibniz. Why does God exist? Why do the Laws exist? Why isn't there just absolute, blank, lawless non-existence? The brute fact remains: Reality is here, and it seems entirely unnecessary.

The Brute Fact

We are left with a few uncomfortable possibilities.

Option A: It is a Brute Fact.

The universe simply is. There is no deeper reason. The chain of explanations has to stop somewhere, and it stops at the laws of physics. Asking "why" beyond that point is a category error, like asking "What is the flavor of the number seven?" It's a question that sounds grammatical but has no meaning.

Option B: Nothing is Impossible.

Perhaps "nothing" is a logical contradiction. Some philosophers argue that a state of absolute non-existence cannot exist because, by definition, there would be no one to verify it, no properties to define it, and no truth to be found in it. Therefore, "something" is the necessary default. Existence is mandatory.

Option C: The Mathematical Universe.

Max Marketer and others argue that mathematics isn't just a description of reality; it is reality. If a mathematical structure (like the universe) is logically possible, then it exists. Since "nothing" corresponds to the null set (which is still a mathematical concept), and since infinite mathematical structures are possible, all of them exist. We are just living in one of the equations.

The Mirror

Ultimately, the question "Why is there something rather than nothing?" might tell us less about the universe and more about ourselves. We are creatures of cause and effect. We evolved in a world where every rock that falls was pushed, and every fire that burns was lit. Our brains are hardwired to look for the "pusher."

When we look at the cosmos and find no pusher only an eternal, vibrating, self-creating void we feel anxious. We feel that something is missing. But the universe is under no obligation to satisfy our need for narrative. The "nothing" we fear might just be the limit of our own imagination, a dark horizon beyond which our primate brains can no longer calculate.

Seventeen

Is Nothing More Natural Than Something?

The Bias of Entropy

When we look at the world around us, we see a universal tendency toward degradation. Structures crumble, living things die, hot coffee cools down, and order dissolves into chaos. This is the Second Law of Thermodynamics: entropy always increases. Because we see complex things constantly breaking down into simpler components, we intuitively assume that the "ultimate" state of simplicity is non-existence. We assume that "nothing" is the baseline, the path of least resistance, the state the universe would naturally slide into if it stopped making an effort.

This creates the "Asymmetry Problem." We view existence as a heavy lift a state that requires energy, design, and maintenance while we view non-existence as effortless. Therefore, the existence of the universe feels like a massive deviation from the norm that demands a specific explanation. We ask, "Why is there something?" because we assume that if things were left to themselves, there would be nothing.

But what if our intuition is backward? What if "nothing" is not the easiest state to achieve, but the hardest? What if "nothing" is actually an incredibly specific, highly ordered, and unnatural state that requires immense effort to maintain?

The Burden of Empty Space

Consider the complexity required to create a perfect vacuum. As we discussed in the first section of this book, creating a region of "nothing" in a lab requires thick steel walls, massive pumps, magnetic shielding, and cooling systems. You have to actively fight against the pressure of the atmosphere, the intrusion of radiation, and the heat of the environment. In the physical world, "emptiness" is not a default; it is a highly artificial condition that nature abhors and tries to destroy.

Now, apply this logic philosophically to the concept of absolute non-existence. To have a true "Nothing" no atoms, no fields, no laws, no space, no time you need a state of perfect constraint. You need a cosmic rule that says: "No particle shall appear here. No quantum fluctuation shall occur here. No geometry shall form here."

Who enforces that rule? A state of absolute nothingness requires a perfect, unbreakable law of prohibition. It requires a "lid" to be kept tightly on the pot of possibility. In this light, "nothing" looks less like a simple void and more like a highly tuned, restrictive state of zero entropy. It is a state of maximum order (because it is perfectly uniform).

"Something," on the other hand, is messy. It is chaotic. It is unconstrained. In quantum mechanics, the rule is often stated as: **"Anything that is not strictly forbidden is mandatory."** If there is no specific law stopping a particle from popping into existence, it eventually will. Therefore, perhaps existence is the true default because it is the state of maximum freedom. "Something" is what happens when you don't have a Supervisor to enforce the emptiness.

The Lawless Void

This leads to a paradox concerning the "Philosopher's Nothing" we discussed in the previous chapter. If we imagine a state *before* the universe where there were no laws of physics no quantum mechanics, no relativity, no conservation of energy then there were also no laws to forbid things from happening.

If there are no laws, there is no law of "Conservation of Energy" to say you can't create matter. There is no law of "Causality" to say effects must have causes. There is no law of "Logic" to say A cannot be non-A.

In a truly lawless void, there is nothing to stop a universe from simply appearing for no reason at all.

This suggests that true, absolute nothingness might be logically unstable. It is so devoid of properties that it lacks even the property of "preventing existence." The moment you remove the laws of physics to get to "pure nothing," you remove the very barriers that keep the universe out. Thus, "nothing" might inevitably collapse into "something" simply because there is no rule to say it shouldn't. The void fills itself because it is too incompetent to stay empty.

The Selection Bias of Reality

Even if we accept that "something" and "nothing" are both possibilities, we still have to explain why *we* are here. Why did the coin flip land on "existence"? This brings us to the **Anthropic Principle**.

Imagine a lottery with a trillion tickets. The odds of you winning are effectively zero. If you win, you will likely scream, cry, and ask, "Why me? Why did I win? There must be a reason!" You might attribute it to fate, karma, or divine intervention.

But to the lottery organizer, there is no mystery. Someone had to win. If enough people buy tickets, a winner is a statistical certainty. The winner asks "Why me?" only because the losers aren't there to ask "Why not me?"

The Anthropic Principle applies this logic to the cosmos. If we live in a Multiverse (as suggested by Inflation and String Theory), there may be trillions of regions of space that are truly empty. There may be aeons of time where nothing happens. There may be universes that collapsed instantly or tore themselves apart.

But those regions have no observers. There is no one in the empty voids to look around and say, "Why is there nothing?"

The only regions that contain questions are the regions that contain matter. We find ourselves in a universe full of stuff stars, planets, carbon, light not because it is the only possibility, but because it is the only possibility *that includes us.* We are suffering from survivor ship bias. We look at the successful universe and assume it is special, when in reality, we are simply the winners of a cosmic lottery that played out billions of times in the dark.

The Necessity of the Question

This reasoning—that we are here because we couldn't be anywhere else often feels unsatisfying. It feels like a tautology: "We exist because we exist." It lacks the emotional weight of a "Grand Plan."

But science does not promise emotional satisfaction; it promises consistency. The realization that "nothing" is a highly constrained state, while "something" is a chaotic probability, shifts our perspective on the value of life. We are not a deviation from a perfect, peaceful void. We are the inevitable result of probability operating on an infinite scale.

If nature abhors a vacuum, then existence is not a miracle; it is a necessity.

We are the exhaust fumes of the void's inability to remain still. We are the realization that "zero" is merely a number, and nature prefers to count.

Eighteen

Why Our Brains Fail at Nothing

The Savanna Operating System

If you have felt dizzy, confused, or skeptical while reading this book, do not worry. You are supposed to. The human brain is the most complex object in the known universe, but it was not designed to understand the universe. It was designed to help a hairless ape survive on the grasslands of East Africa.

Our neural architecture evolved over millions of years under specific pressures: find food, find a mate, avoid predators, and navigate social hierarchies. To succeed in these tasks, our brains developed a set of intuitive physics engines. We instinctively understand that:

- Solid objects cannot pass through each other.
- Things do not move unless pushed.
- If you can't see a lion in the grass, it might still be there (but the space between the grass is irrelevant).
- "Nothing" means "no threats" or "no food."

We are wired to process "medium-sized dry goods" rocks, trees, animals moving at medium speeds. We are not wired to process the quantum vacuum, curved spacetime, or 11-dimensional strings. When we try to visualize "empty space that has weight" or "particles that are also waves," we are running high-end software on hardware that was built to throw spears. We are forcing the Savanna Operating System to run a simulation of the cosmos, and it is inevitable that the system will crash.

The Horror of Non-Existence

Our primary cognitive limitation regarding "nothing" is that our imagination is additive. Try to imagine a unicorn. Your brain assembles a horse and a horn. Now, try to imagine *no* unicorn. You don't imagine a "void of unicorns"; you simply imagine a horse-less pasture or a blank room. We visualize "nothing" by visualizing "something else" in its place.

This is because our visual cortex is an object-recognition machine. It scans the environment for edges, contrasts, and shapes. A true void has no edges, no contrast, and no shape. To the brain, it is invisible data.

Consider the "Blind Spot" in your eye. There is a patch on your retina where the optic nerve connects, which has no light receptors. You technically have a hole in your vision. But you never see a black patch. Your brain invents "filler" data it copies the pattern from the surrounding area and pastes it over the hole to create a seamless illusion of completeness. We do the same thing conceptually. When we think of the vacuum of space, our brains paste "empty air" or "dark canvas" over the concept. We cannot mentally represent a true lack of properties, so we unknowingly fill the void with imaginary properties just to keep our worldview coherent.

The Agency Trap

Another evolutionary glitch that hinders our understanding of the universe's origin is our **Hyper-Active Agency Detection Device (HADD)**.

Imagine you are a prehistoric human walking past a bush, and you hear a rustle.

- **Mistake Type 1:** You assume it's the wind, but it's a tiger. Result: You die.
- **Mistake Type 2:** You assume it's a tiger, but it's the wind. Result: You get scared, but you live.

Because the cost of Mistake Type 1 is death, evolution selected for paranoia. We evolved to assume that *everything* that happens is caused by an agent a conscious being with intent. If the grass moves, an animal moved it. If thunder crashes, a god threw it. If the sun rises, a chariot pulled it.

This cognitive bias makes the idea of a "Universe from Nothing" deeply unsatisfying. When we ask, "How did the universe begin?", our brains are actually asking, "Who started it?" We are looking for a cause that looks like a person (a Creator, a Watchmaker). The answer offered by quantum mechanics that the universe is a spontaneous fluctuation of a probabilistic field feels "wrong" because it lacks an agent. It feels like a rustling bush with no tiger. Our intuition screams, "That's impossible! Things don't just happen!" But the universe is under no obligation to respect our paranoia.

The Problem of Zero

We also struggle with the mathematical concept of zero, as discussed in Chapter 2, because we tend to think of numbers as adjectives, not nouns. To our intuition, "three" describes "three apples." "Zero" describes "zero apples." The apples are the real thing; the number is just a label.

But in modern physics, the numbers *are* the real thing. The fields that make up the vacuum are mathematical entities. When we say the vacuum energy is non-zero, we aren't counting invisible apples; we are describing the intensity of a field.

We have a "folk physics" that treats space as a container and matter as the content. We intuitively feel that you can take the content out and leave the container behind. But General Relativity tells us that the container (spacetime) and the content (energy) are the same thing. You cannot have the "where" without the "what." This destroys our intuitive separation of "Something" and "Nothing." To the universe, they are just different values of the same variable.

The Metaphor Ceiling

Finally, we are trapped by our language. All human language is built on metaphors derived from physical experience. We "grasp" ideas. We "see" the point. Arguments are "solid" or "shaky."

When physicists describe the vacuum, they are forced to use metaphors that are fundamentally wrong because we have no words for the reality.

- They call it a "sea" of particles (but it's not wet).
- They call it "foam" (but it's not bubbly).
- They call it a "field" (but there's no grass).
- They say particles "spin" (but they don't actually rotate).

These metaphors give us a false sense of understanding. We picture a bubbling ocean and think, "Aha, I get it." But we don't. The math describes a reality that is completely alien to our sensory experience. We are like dogs trying to understand algebra; we might recognize the tone of voice, but the meaning is forever beyond our reach.

The Humble Ape

Recognizing these cognitive failures is not a defeat; it is a liberation. It allows us to let go of the demand that the universe "make sense" in a human way.

We must accept that our feeling of "intuition" is a survival tool, not a truth-finding tool. When quantum mechanics says a particle can be in two places at once, or that empty space weighs a billion tons, our intuition shouts "No!" We must learn to ignore that shout. We must trust the equations more than we trust our own brains.

The fact that we creatures of mud and savanna have managed to discover the Higgs Boson, Dark Energy, and the Big Bang is a miracle of abstraction. We have used mathematics to build a ladder out of our own cognitive limitations. We have peered over the edge of our evolutionary blinders and glimpsed the true, terrifying, beautiful nothingness that lies beyond. We may never truly *grok* it, but we can prove it is there.

Nineteen

What Nothingness Teaches Us About Everything

The Great Inversion

We began this book with a simple, intuitive picture of the world: a universe made of solid objects moving through empty space. We thought of "things" stars, atoms, people as the primary actors of reality, and "space" as the passive stage on which they stood. But after traveling through the atomic nucleus, diving into the quantum vacuum, and witnessing the accelerating expansion of the cosmos, that picture has been completely inverted.

The greatest lesson of modern physics is that the "nothing" is the primary reality, and the "something" is the afterthought.

We now know that the vacuum is the fundamental substance of the universe. It is a vibrant, energetic, omnipresent ocean of quantum fields. What we call "matter" the protons, neutrons, and electrons that make up your body are merely temporary excitation of this ocean. They are waves on the surface of the deep. To believe that the waves are real but the ocean is "nothing" is a profound error of perspective. We are not distinct objects moving *through* the void; we are patterns *of* the void.

This shift from an "object-oriented" view of reality to a "field-oriented" view changes everything about how we understand our place in the cosmos. It dissolves the rigid boundaries between "self" and "other," between "here" and "there." If you are made of the same fields that stretch to the edge of the visible universe, then you are continuous with the stars. The separation is an illusion created by the stability of the waves.

The Generosity of Zero

The second great lesson is that the universe is inherently generous. Classical logic told us that you cannot get something for nothing. It told us that zero is a dead end. But quantum mechanics has shown us that zero is a seed.

We learned that the vacuum is restless. It cannot remain still. It seethes with virtual particles, borrowing energy and paying it back in a constant cycle of creation and destruction. We learned that the universe itself likely sprang from such a fluctuation a "free lunch" where the positive energy of matter was balanced by the negative energy of gravity.

This tells us that existence is not a rare, difficult miracle that requires external intervention. Rather, existence is the inevitable consequence of the laws of physics. The void is fertile. It wants to be full. The fact that we are here is not a stroke of impossible luck, but a testament to the fact that "nothing" is unstable. In a sense, the universe exists because it is too energetic to not exist.

The Precariousness of Being

However, this generosity comes with a price: fragility. We have learned that our reality is built on a foundation that might not be as solid as we thought.

We discovered that the "solid" matter of our daily lives is 99.9999999% empty space, held together by forces that are transmitted by virtual particles. We discovered that the expansion of the universe is accelerating, driven by a Dark Energy that will eventually isolate every galaxy in a cold, lonely dark. And we faced the terrifying possibility of Vacuum Decay the idea that our universe might be resting on a "false floor," waiting for a quantum tunnel event to rewrite the laws of physics and erase us in an instant.

This knowledge instills a deep sense of cosmic humility. We are not the masters of a static, eternal kingdom. We are campers pitching a tent on the side of an active volcano. Our existence is a temporary equilibrium, a lucky balance of forces that allows for complexity to arise for a brief window of time between the Big Bang and the eternal freeze. We should treasure this complexity precisely because it is fragile. We are the rare, thinking foam on top of a mindless, churning sea.

The End of Certainty

Perhaps the most difficult lesson is the death of certainty. Heisenberg showed us that we cannot know the position and momentum of a particle simultaneously. Goedel showed us that mathematics has limits. Cosmology has shown us that 95% of the universe (Dark Energy and Dark Matter) is made of stuff we do not understand.

We have to accept that our "common sense" is a local adaptation, not a universal truth. The more we learn about the nothingness, the more we realize how strange reality truly is. It allows for particles to be in two places at once. It allows for space to bend and stretch. It allows for energy to appear from nowhere.

This should not lead to despair, but to curiosity. The fact that the universe is weirder than we *can* imagine means that the age of discovery is far from over. The mystery of the vacuum energy the 10^{120} discrepancy is not a failure of science; it is an invitation. It is a signpost pointing toward a new, deeper theory of gravity and quantum mechanics that we have yet to uncover. The nothingness is guarding the greatest secrets left to find.

The Voice of the Void

Finally, what does this mean for us, the conscious observers?

We are the part of the vacuum that has woken up. For billions of years, the universe was a play of fields and forces, a silent ballet of energy. But in the intricate arrangement of neurons in the human brain, the vacuum became complex enough to ask questions. It became complex enough to wonder, "Why am I here?"

When you look into the dark night sky, you are not looking at an alien abyss. You are looking at your ancestor. You are looking at the potential from which you emerged and the silence to which you will return. We are distinct from the void not because we are made of different stuff, but because we are the void singing. We are the mechanism by which the nothingness knows itself.

This realization is the ultimate antidote to the horror vacuity the fear of the empty space. We do not need to fear the nothing, because we *are* the nothing. We are the fleeting, beautiful, conscious shape that the nothing takes for a little while. And that is enough.

This concludes the main chapters of "THE SCIENCE OF NOTHINGNESS."

Twenty

EPILOGUE

THE FULLNESS OF THE VOID

The Room Revisited

We began this journey with a thought experiment. In the Prologue, I asked you to imagine the room you are sitting in. I asked you to mentally strip away the furniture, the air, the light, and the atoms, until you were left with a perfect, empty void. At the start of this book, that exercise likely felt like an act of destruction a stripping away of reality until only a blank, dark canvas remained. You probably imagined a silence so profound it was terrifying, a stillness that signified the absence of existence.

But now, as you close these pages, I invite you to perform that experiment one last time. Look at the space in front of your eyes. Clear it out again. Remove the air. Remove the radiation. Remove the matter.

What do you see now?

You no longer see a blank canvas. You see a frantic, boiling ocean. You know that in that cubic meter of "emptiness," trillions of virtual particles are flickering in and out of existence every nanosecond, borrowing energy from the bank of the universe and paying it back in a flash of annihilation. You know that the space is heavy that it carries the weight of Dark Energy, a mysterious pressure that is currently pushing the galaxies apart and stretching the fabric of the cosmos. You know that if you were to place two metal plates there, the sheer force of this nothingness would crush them together.

You know that the "silence" you imagined is actually a cacophony of quantum fields the electromagnetic field, the electron field, the Higgs field all vibrating at their lowest frequencies, humming the baseline tune of reality. You know that this vacuum is not just a container, but a shoplifter; it can curve under the influence of gravity, it can tunnel into new states of physics, and it once blew up from the size of a proton to the size of a universe in the blink of an eye.

The Paradox of Presence

We spent this entire book looking for nothing. We chased it down the microscope to the atom, and out the telescope to the edge of the Big Bang. And the final, ironic discovery is that we failed. We never found "nothing." Every time we thought we had reached the bottom of the bucket, we found another layer of "something" underneath.

We found that "empty" space is a material. We found that "zero" is a balance of infinities. We found that "absence" is just a specific type of presence.

This failure is the greatest triumph of physics. It teaches us that existence is sticky. You cannot scrape it off the windshield of reality. The universe is incapable of being truly empty because the laws of physics themselves are a form of presence. As long as the potential for something exists, the void will fill itself. The distinction between "being" and "non-being" turns out to be a blurry line drawn by human language, not by nature.

The Void Singing

So, do not fear the emptiness. We tend to view the void as a symbol of death, of loneliness, of the final end. But science tells a different story. The void is the womb. It is the reservoir of infinite potential from which all particles borrow their mass and to which they all return.

You are not an object isolated in a hostile vacuum. You are a localized vibration of the vacuum itself. You are a temporary knot in the fields that fill the universe. When you look up at the dark sky, you are not looking at a stranger. You are looking at the silent, darker part of yourself.

We are distinct from the void not because we are something different, but because we are the void singing. For a brief moment, the fields have tied themselves into a knot capable of reading a book, thinking a thought, and feeling wonder. And when that knot eventually unties, the energy that made us will simply return to the ocean, ready to rise again in a new wave, in a new time, in a new universe.

The room is empty. The room is full. Both are true. And that is the miracle.

www.ingramcontent.com/pod-product-compliance
Lightning Source LLC
LaVergne TN
LVHW090421210926
848480LV00002B/646